Thomas Naff, Series Editor

The Euphrates River and the Southeast Anatolia Development Project

John F. Kolars

William A. Mitchell

Southern Illinois University Press

Carbondale and Edwardsville

Library of Congress Cataloging-in-Publication Data

Kolars, John F.
    The Euphrates River and the southeast Anatolia development project
/ John F. Kolars, William A. Mitchell.
        p.   cm. — (Water: the Middle East imperative)
    Includes bibliographical references and index.
        1. Water resources development—Turkey.   2. Water resources
    development—Euphrates River Watershed.   3. Stream measurements—
    Syria.   4. Stream measurements—Iraq.   I. Mitchell, William A.,
    1940–   . II. Title. III. Series.
    TC512.E8K65   1991   91-10879
    333.91′15′09561—dc20   CIP
    ISBN 0-8093-1572-6

To Dr. F. Otto Haas,
*without whose enlightened support our work
would not have been possible*

Contents

Maps

Figures

Tables

xviii

Abbreviations and Acronyms

B.C.E.	Before Christian Era
DSI	Devlet Su Isleri (The General Directorate of State Water Works)
FAO	Food and Agriculture Organization (of the United Nations)
GADEB	General Authority for the Development of the Euphrates Basin
GAP	Guneydogu Anadolu Projesi (Southeast Anatolia Development Project)
GNP	Gross National Product
HEPP	hydroelectric power plant
HES	hydroelectric station
IBRD	International Bank for Reconstruction and Development
PE	potential evapotranspiration
RF	return flow
RPU	Resource Planning Unit
SAR	Syrian Arab Republic
TEK	Electrical Authority
TCSV	Turkiye Cevre Sorunlari Vakfi (Turkish Environmental Issues Foundation)
TL	Turkish lira

TS	Toprak Su (General Directorate of Land and Water Development—no longer in operation)
TTRM	Toprak ve Tarim Reform Mustesarligi (Secretariat for Soil and Agricultural Reform)
TUBITAK	Turkiye Bilimsel ve Teknik Arastirma (Turkish Foundation for Scientific and Technical Research)
UAR	United Arab Republic
USGS	United States Geologic Survey
USAID	United States Agency for International Development

Note on Transliteration

In rendering proper names from Middle Eastern languages into English, two standards have been used.

For Turkish, the names have been given as they are spelled in modern Turkish, with European characters but minus the diacritics. This is an aid to the reader who may find these names on modern Turkish maps.

For Arabic and other languages that utilize non-European characters, the nomenclature adheres as closely as possible to current usage as found in reputable nonspecialist journals in the English language, again without diacritics or special symbols. Our reason for this is simple: The expert doesn't need these aids; the nonexpert cannot use them.

Editor's Preface

This volume, which concentrates on the Euphrates River basin in Turkey and Syria, is the first in a series on water issues in the Middle East. In it, Professors John Kolars and William Mitchell have sketched out the impressive water developments now under way in the upper reaches of the Euphrates River in Turkey. They then go on to apply sophisticated methods of analysis in order to estimate what impact these developments will have on the downstream riparians Syria and Iraq, who also, both historically and currently, have been engaged in hydraulic developments.

The complete water project, of which this present work forms a part, consists of comprehensive assessments of several of the most important hydrological basins in the Middle East: the Euphrates, Tigris, Jordan, Litani, Orontes, and Nile. Work has been conducted under the auspices of Associates for Middle East Research, Inc. (AMER), a non-profit research group in Philadelphia, and has been directed by Professor Thomas Naff of the University of Pennsylvania. The project began in 1983 with a broad-based pilot survey which was published the following year under the title *Water in the Middle East: Conflict or Cooperation?* (Westview Press, Boulder, Colorado, 1984). This earliest phase of the research identified the major issues and demonstrated the need for far more detailed analysis than was possible in one slim volume collated from the available literature.

The purview of the present study, and, indeed, of the overall project, is both physical and political, including socioeconomic factors. Thus, each geographic segment is looked at twice; first, as part of its hydrologic basin to determine the physical dimensions of its water problems, then as part of the country or countries within which it falls to show the contribution of water problems to the political economy and strategic concerns of its region. The end product is an illustration of the inter-relationship between natural resources and international security.

The present volume is the first installment of a larger effort to address those Middle East water issues that the world and the peoples of the region itself can no longer afford to ignore.

The project from which the book has grown has brought together a team of specialists from the Middle East, Europe, the United States, and Australia, drawn from such disciplines as hydrology, geography, geology, engineering, history, politics, economics, sociology, demography, international relations, international law, and resources management. The work is based on extensive field data as well as information from published sources.

The results of this interdisciplinary research will be published over the next three to five years as a series of books. The present volume, on the hydrology of the upper Euphrates in Turkey and Syria, will be followed by hydrological analyses of the lower Euphrates and of the Tigris basin as they interconnect in Iraq. Future volumes will discuss the hydrology of the Jordan and Litani basins and their relationships, and the current hydrological situation in the Nile basin.

Companion volumes will examine the political economy and apparatus for water policy decision making in Turkey, Syria, Jordan, Iraq, Israel, Lebanon, Egypt, and Sudan. These studies will give special focus to the strategic and security implications of water problems for each of the countries and for the region as a whole. The legal status of water regulation in the area will be investigated in a separate volume, which will present a compendium and commentary on the relevant treaties, legal codes, and regulatory systems that govern water management and water sharing in the Middle East. Another volume that reviews current and known future technologies will examine the extent to which technology can play a mitigating role in water problems.

Water is a resource vital to life. As any archaeologist looking for prehistoric habitation can testify, a reliable water source is the one absolute prerequisite for human settlement. It is the ultimate survival issue, a "superordinate goal" that overrides all other concerns. With populations expanding and aspirations for economic development increasing, the demand of Middle Eastern peoples on their limited water resources is already approaching the "water barrier" beyond which the need for water becomes a dominant concern. Moreover, overdrafts on existing water resources can degrade remaining supplies, reducing quality below usable levels, and actually decrease the annual amount of water available for future human consumption. In the seven years since this project began, it has increasingly been realized that the most important Middle East resource issue of the twenty-first century will be, not oil, but water.

Restricted supply of water in the Middle East has been a truism since ancient times. Scarcity, that is, the point at which water demand and its related quality of life are constrained by the inadequacy of the available supply, is already a fact in some sectors. Several countries are running a significant deficit on their water budgets which, at best, constitutes a lien against future generations and may, in the worst case scenario, destroy the existing resources for future use. The need to conserve on use of water and somehow augment supplies is very clear.

The threat to posterity and even survival that water scarcity represents is an automatic security issue for all parties. It is rooted in real human needs. When a limited water resource is shared by two nations, competition for it can be a cause of tension; when these nations are already hostile, the competition could lead to open conflict. On the other hand, the "superordinate goal" of mutual survival may compel even hostile neighbors to cooperate. Therein lies hope for agreement, but the opportunity for avoiding a regional crisis is evaporating rapidly.

There are relatively few treaties or agreed-upon international legal or political structures for settling disputes over shared resources. In this unregulated environment, conflicts of interests tend to become confrontations to be dealt with in terms of regional and global power balances. It is our belief that a holistic understanding of resource problems, including their security implications, will empower countries with options for managing their environments without recourse to conflict. This in turn will enhance the capacity of the international system to create legal and other mechanisms for the resolution of problems through accommodation.

The study of water, like other resource issues, is complex, requiring a wide range of expertise from a variety of disciplines from the hard sciences and engineering through the whole gamut of social and managerial sciences. The result of this complexity has, in the case of water, been fragmentary treatment at best, and often outright neglect.

The magnitude of the water problems and the interconnections of the water systems dictate that, for the Levantine, Anatolian, and Mesopotamian areas we are studying, only truly regional solutions will work. The essential first step in that direction is basin wide agreements. This series of publications constitutes our contribution to making such regional cooperation possible.

As with all weather-related phenomena these days, a significant unknown in the calculus that may throw all prognostications off is the so-called Greenhouse Effect, which will surely make Middle East water problems worse. It behooves planners to address the issues now, while solutions are still at least theoretically within reach.

A project of this complexity and magnitude could not have been successfully completed without the cooperation, support, and assistance of many persons, institutions, funding agencies, and governments. The Director and all of his colleagues express their gratitude to all those individuals, too numerous to list here, whose assistance was pivotal in bringing our efforts to fruition. We also wish to extend our sincerest thanks to the government of Turkey and various of its agencies who gave us enlightened access to essential data, and to the following sponsors whose contribution to our project made it all possible: The Phoebe W. Haas Foundation, Ford Foundation, Mobil Foundation, the University of Pennsylvania Research Foundation, Bechtel Corporation, Conoco Oil Company, Dow Chemical, Dresser Industries, E. I. Du Pont de Nemours and Company, Exxon Corporation, and Fluor Corporation.

THOMAS NAFF
PHILADELPHIA

The Euphrates River and the Southeast Anatolia Development Project

1

Turkey, the Euphrates-Tigris River Basin, and the Southeast Anatolia Project

History has been said to begin at Sumer, and history today continues to be made in the combined basins of the Euphrates and Tigris rivers. Increasing water shortages in southwest Asia, in combination with the ambitious development plans of every nation found there, focus attention upon those two rivers. Constituting the region's major sources of water, their proper management in the years ahead will help determine the welfare and political stability of much of the Middle East.

Turkey occupies the position farthest upstream on both rivers. Almost all the waters of the Euphrates and a major portion of the waters of the Tigris come from within Turkey's borders. Unlike many Middle Eastern countries, Turkey is petroleum poor but water rich. The nation receives about 509 billion m^3 of precipitation annually, of which 38 percent (185 billion m^3) ends up as surface runoff. Much of this flows into the USSR, Iraq, Iran, Syria, and the surrounding seas. Because the Turks estimate that only a little over half of this surface runoff (95 out of 185 billion m^3) can be used for domestic, irrigation, and industrial purposes within Turkey, the international implications of the situation are obvious (DSI 1984b, Tables 4 and 4.1, 20).

Turkey is under enormous pressure to develop its hydroresources (Kolars 1986a). Total energy use in Turkey from 1975 to

1982 increased by 30 percent, while energy from all Turkish sources increased by only 24 percent. In 1983 nearly 39 percent of the energy used in Turkey came from petroleum imports. If imports of coal and electricity are added to the ledger, 40 percent of all the nation's energy came from beyond its borders. Before the drop in world oil prices Turkey paid over $4 billion a year for imported petroleum products. Even with the subsequent drop in crude prices, over $2 billion are currently spent on imported oil. In 1985 about one-tenth of the Turkish energy base was provided by hydropower, and it is clear that future substitution of additional hydropower for thermal energy will save much needed exchange credits. Turkey further hopes to balance its import-export ratio by selling large quantities of agricultural produce to its Arab neighbors to the south (Kolars 1986b). Paramount among the sources of such produce anticipated by Turkish planners will be the vast new irrigated fields of the Southeast Anatolia Project.

In consequence, the nation is rapidly developing its rivers through the construction of dams, reservoirs, hydropower plants, and irrigation projects. By 1987 the General Directorate of State Water Works (Devlet Su Isleri, or DSI) managed a national construction program including 134 major dams, 72 hydroelectric power plants, 158 smaller irrigation and detention dams, and 521 groundwater projects. Hydroelectric power from these and other proposed projects will eventually provide 53 percent of all electricity consumed in Turkey. Once completed, this national development scheme will irrigate 5,925,032 ha of land, convey 2,520.1 Mm^3 of water to settlements, and produce annually 110.117 million kWh of energy from 426 hydropower plants.

Central to all this is the Southeast Anatolia Development Project (Guneydogu Anadolu Projesi, or GAP), which is the subject of this study. The importance of GAP can scarcely be overstated. Turkish estimates based on complete development of the nation's water resources show that the twin basins of the Euphrates and Tigris rivers within Turkey account for 29.9 percent of the country's total surface runoff. This in turn could provide water to irrigate 2,032,203 ha (34.3 percent of all possible irrigated land) as well as 43.5 percent of projected hydropower from all such projects (DSI 1984b, 27).

Turkey's downstream riparian partners on the Euphrates and Tigris rivers, Syria and Iraq, also have much at stake. They, too, are becoming increasingly dependent on the finite supply of water provided by the two rivers. Electric generation from the generators at the Tabqa Dam[1] on the Syrian Euphrates in 1979 amounted to 60 percent of all the electric energy produced that year in Syria. Prior to the construction of the dam, 180,000 ha were irrigated in the valley of the Syrian Euphrates; as much as 650,000 ha were to be brought under irrigation after the dam's completion. Although these plans now seem overly ambitious—as will be seen in the chapters that follow—Syria still places high hopes and great dependence upon its use of the Euphrates River (Meliczek 1987).

Downstream, the third riparian, Iraq, is engaged in development not only of the Euphrates but also of the Tigris and its tributaries. Over 1 million ha of Iraqi land currently receive water from the Euphrates. Although this study does not include an analysis of Iraq's use of these waters, the urgency of its situation—which will be described in another publication in this series—is just as great as that of its upstream neighbors.

Given the dimensions of this situation and its scope in time and space, it will be useful—before considering the technical details on which political negotiations among users of these waters will depend—to take a brief historical and regional view of this strategic area.

The Euphrates (Firat) River

The Euphrates River has its sources in eastern Turkey and its mouth at the head of the Persian/Arabian Gulf. Along with the Tigris and Karun rivers, the Euphrates brings water to the Mesopotamian lowlands of Iraq as well as hydropower and irrigation to parts of southeastern Turkey and much of northern and eastern Syria. It is the longest river (2,700 km) in southwest Asia west of the Indus, although its maximum average annual volume (35.9 billion m^3 at Hit, Iraq) is less than that of the Tigris (70.4 billion

---

1. The Tabqa Dam is now formally referred to as al-Thawra.

m³ at Baghdad) or the Karun (48.8 billion m³ at Ahwaz) (Cressey 1958). Because its waters come from melting snows, maximum flows are in April/May, while minimum flows are in September/October.

The river is formed in Turkey by two tributaries: the Karasu which originates at an elevation of 2,744 m north of the city of Erzerum, and the Murat which begins on the slopes of Ala Dag (Mount Ala) north of Lake Van at 3,135 m elevation. The two streams join 45 km northwest of the city of Elazig. Thereafter, the combined Euphrates cuts through the southeastern Taurus Mountains and crosses into Syria at Karkamis (Carablus, ancient Carchemish) downstream from the Turkish town of Birecik. In this portion of its journey the river drops 2 m/km. After Birecik, the river flows southeast across the Syrian tableland in an entrenched valley, where its major tributaries are the Balikh and the Khabur, both of which enter the left bank from the northeast.

At 360 km downstream from the Iraqi-Syrian border, the river reaches its alluvial delta near Hit in Iraq. This is still 735 km from the Gulf, but the river is only 53 m above sea level. From this point on, the river loses part of its waters into a series of desert depressions and distributaries, both natural and man-made. Farther downstream near Nasiriya, the river becomes a tangle of channels, some of which drain into shallow Lake Hammar while the remainder join the Tigris River near Qurna. From Qurna to the sea the combined streams are known as the Shatt al-Arab. The Shatt is joined 32 km below Basra by the Karun River, which flows westward from the Zagros Mountains of Iran. The combined flow continues another 77 km before entering the Gulf.

Navigation on the Euphrates has traditionally been confined in its upper reaches to brushwood rafts (*kellek*) supported by as many as 800 inflated sheepskins. In parts of Syria and Iraq, traditional craft have included coracles (*quffas*) and sailboats (*mahaila, safina*). Upstream from Hit, rapids and shoals bar modern transport; downstream shallows and shifting sandbars do the same. Thus, despite repeated attempts to maintain channels, steamer navigation has not been possible although the government of Iraq anticipates improving navigation on the river in the future. The Shatt al-Arab is navigable by small steamers as far as Basra, where further limited steamer traffic is possible as far as Baghdad and Mosul on the Tigris.

Historically, the Euphrates derives its name from the Sumerian *Buranun*, which became *Purattum* in Akkadian, *Ufrat* in Old Persian, *Euphrates* in Greek and Latin, *Furat* in Arabic, and *Firat* in Turkish. Nippur and Ur (both Sumerian cities) and Babylon were near or on the river, as was Carchemish, all of which date from before the first millennium B.C.E. While the river in ancient as in recent times offered some means of transportation, most of the traffic was in a downstream direction. Rapids and shallows made the upstream journey too difficult. Towns along the river thus became way stations on the east-west caravan routes which crossed the region, arching northward toward the highlands of present-day Turkey.

The main contribution of the Euphrates to the ancients was as a source of irrigation water. Water was and still is lifted by giant water wheels called *norias* onto the fields in Syria and northern Iraq. Water was also brought to the land by canals, made possible because the Euphrates is higher than the Tigris north of Baghdad but lower than the Tigris south of that city. Thus, irrigation water from one river can be drained away into the other stream and into canals paralleling the two rivers. In modern times, gasoline and electric pumps play a significant role in lifting water.

As early as the fourth millennium B.C.E. agricultural settlements with temples and local irrigation networks were part of the Mesopotamian landscape. The Sumerians and Babylonians brought water to their fields and cities by canals from the Euphrates. Documents from the time of Hammurabi, the Babylonian law-giver of the early second millennium, refer to maintaining the irrigation systems. Similar irrigation works were undertaken by rulers as late as the Abbasid Caliphs (A.D. 750–1258).

The destruction of much of the canal system during the Mongol invasion of the thirteenth century, combined with subsequent neglect and the breakdown of central administration, led to a general abandonment of such works until modern times. Poor drainage in the lowlands near the river allowed salts from the mineral-rich irrigation water to poison the soil. Beginning in the nineteenth century, much of this land has been gradually reclaimed through careful washing of the salts from the fields. Ancient canals have been cleaned and rebuilt, and new systems constructed for irrigation and drainage. Barrages such as the one at Hindiya in Iraq (built in 1913) have been erected across the river to raise its level and divert water for additional irrigation. Large dams and reservoirs have been and are being constructed in Turkey and Syria. It is the purpose of this study to place the projects of those two countries in clearer perspective.

The Tigris River

The Tigris River is the second longest river in southwest Asia (1,840 km). Its name comes from the Sumerian *Idigna*, which became *Idiglat* in Akkadian, *Tigra* in Old Persian, and *Tigres* to Herodotus (*circa* 450 B.C.E.) and those after him. Modern Turks refer to it as the *Dicle*, pronounced "Dijlay," which is also the Arabic name.

The Tigris rises in eastern Turkey near Lake Hazar (elevation 1,150 m) and flows southeast to the Turkish city of Cizre, where it forms the border between Turkey and Syria for 32 km before entering Iraq. Midway between Tikrit and Samarra in Iraq the river enters its delta, and from there on forms the eastern part of the complex Tigris-Euphrates system, which both waters and drains the lowlands of Mesopotamia. On its journey through Iraq numerous tributaries enter the left bank of the Tigris from the Zagros Mountains to the east. Among these tributaries are the Greater Zab, the Lesser Zab, the Adhaim, and the Diyala. Near Qurna in southern Iraq, the Tigris and Euphrates join and continue as the Shatt al-Arab for the remaining 179 km to the Gulf.

Great swamps stretch on both sides of the Tigris from Qurna northward for 80 km. The river is navigable with difficulty as far north as Baghdad. From Baghdad to Mosul the route is sometimes plied by shallow draft motorboats, but above Mosul any river traffic is downstream by *kellek* rafts only.

The location of the Tigris near the foot of the Zagros Mountains allows its tributaries to be filled with water from both melting snows in the spring and rainfall in warmer weather. When rain and snowmelt occur together, the accumulation of water is enormous. Thus, the Greater Zab may supply up to 65 percent of the flow of the Tigris for brief periods in April and May. However, so much water is removed by irrigation canals—such as the Hilla and Gharraf canals—and natural distributaries leading into Lake Hammar that the peak flows fluctuate between 14,000 m^3/s near the confluence with the Diyala to a scant 179 m^3/s at Qurna. The volume of the river also varies widely from time to time at any given place. At Baghdad, the minimum recorded flow was 158 m^3/s, the mean is 1,236 m^3/s, and the maximum was 13,000 m^3/s. This compares with a minimum flow of 181 m^3/s and a maximum of 5,200 m^3/s on the Euphrates at Hit.[2] The Euphrates thus has a more regular regime because of its greater length and fewer downstream tributaries; the Tigris is famous for its floods.

The danger of flooding at Baghdad has resulted in a scheme to divert excess water from the Tigris into the Tharthar Depression between the two rivers, and then to release the stored water into either stream as need dictates. A second project, yet to be completed, will provide an outfall drainage canal to remove irrigation water from 1.5 million ha between the Tigris and Euphrates rivers into the Gulf (MEED 1987, 31:12).

Along with the Euphrates, the Tigris has played an important role in human affairs since ancient times. The Assyrian capitals of Asshur, Nimrud, and Ninevah were located on its banks, as were the Seleucid cities of Seleucia and Ctesiphon. From A.D. 197 to 237, the Tigris north of Mosul served as the boundary of Roman Mesopotamia. Later, the Abbasid towns of Baghdad, Samarra, and Mosul prospered there. While these cities took part in the trade and transportation activities associated with the river, their actual prosperity was based on the surrounding irrigated farmlands.

2. Comparison of these figures with those given at the beginning of the previous section demonstrates the complexity of the data pertinent to these discussions. The earlier figures refer to total annual flow, which may vary throughout the year; the latter refer to ephemeral flows which last for varying, and sometimes very brief, periods of time. All such data presented in these introductory comments must be taken as provisional. The detailed analysis in the chapters that follow will give the most accurate and definitive account possible of natural flows and human utilizations.

Modern development of the river is under way. In Iraq, a $1 billion Mosul multipurpose hydroelectric and irrigation project is being implemented. There are existing dams and barrages such as the Darich Dam near the Turkish border, the Aski Mosul and Ba'iji Dams in mid-course, and farther downstream barrages at Samarra and Kut. Other projects including the Bekme Dam on the Greater Zab and the Himrin Dam on the Diyala are planned for the east bank tributaries. In Turkey, the Tigris River is an integral part of the GAP scheme. The Kiralkizi and Dicle Dams are under construction; the design for the Ilisu Dam has been completed; and plans for the Cizre Dam are nearing completion (DSI 1984a).

Turkish National Hydrodevelopment Programs

Because Turkey controls the headwaters of the Tigris and Euphrates rivers, an understanding of the Turkish hydroelectric development program is essential to the study of the two rivers. The first question is, Has Turkey the technological and management skills necessary to accomplish the ambitious plans alluded to above? A brief historical summary of water utilization in modern Turkey, and in the Ottoman and earlier states that preceded the Republic, should provide a firm "yes."

Pre-Republican Times

Construction of water use systems in Asia Minor dates from the Hittite and Urartu periods (as early as 3000 B.C.E.), and has continued almost uninterrupted to the present. Some of the dams built in Van province over a thousand years ago are still in use. People in the Urartu period were pioneers in building small dams and diversions throughout eastern Anatolia. Unlike the large-scale irrigation and flood control systems in other cultural hearths, such as Egypt, Mesopotamia, the Indus and Hoangho valleys, the ancient systems in Turkey were mainly urban water works (Ozis 1987).

In more recent centuries, sophisticated aqueducts and cisterns were built in Asia by Greeks, Persians, Romans, and Turks. Both Seljuk and Ottoman Turks constructed hundreds of foun-

Table 1.1. Dams Constructed in Turkey Prior to the Republic Era

Dam	Location	Reservoir Volume (in m³)	Height (in m)	Year	Builder
Topuz	Istanbul	70,000	8.60	1620	Sultan Osman II
Buyuk	Istanbul	1,318,000	12.15	1724	Sultan Ahmet III
Topuzlu	Istanbul	160,000	16.00	1750	Sultan Mahmut I
Ayvat	Istanbul	156,000	13.45	1765	Sultan Mustafa III
Valide	Istanbul	255,000	13.50	1796	Sultan Selim III
Kirazli	Istanbul	103,080	13.00	1818	Sultan Mahmut II
Yeni	Istanbul	217,500	17.00	1839	Sultan Mahmut II
Elmali	Istanbul	1,700,000	19.75	1893	Sultan Abdul Hamid II

Source: DSI (1984a).

tains and diversion dams for domestic use in towns and villages and for watering animals on common pastures. Table 1.1 lists eight of the largest such dams built by the Ottomans prior to the Republic. They also developed fountains and irrigation projects along caravan routes. The law was clear to all: Water was Allah's gift and no one had an ownership claim to it. Such may not be the case in Turkey today.

DSI

Turkey's hydrodevelopment program is managed by the DSI. Now under the Ministry of Energy and Natural Resources, this Directorate was established by Law Number 6200 on 18 December, 1953. It is charged with "multiple utilization of ground and surface water and [with] the prevention of soil erosion and flood damages." The specific responsibilities of DSI are to build protective structures against floods, to drain swamps, to build irrigation and drainage systems, to construct hydroelectric power generation plants, to develop all stages of water supply and water treatment plants for settlements over 100,000 population, and to improve navigable rivers (although these are almost non-existent in Turkey).

DSI is a three-tiered line organization. Its top management level is the General Directorate office in Ankara (Figure 1.1). The

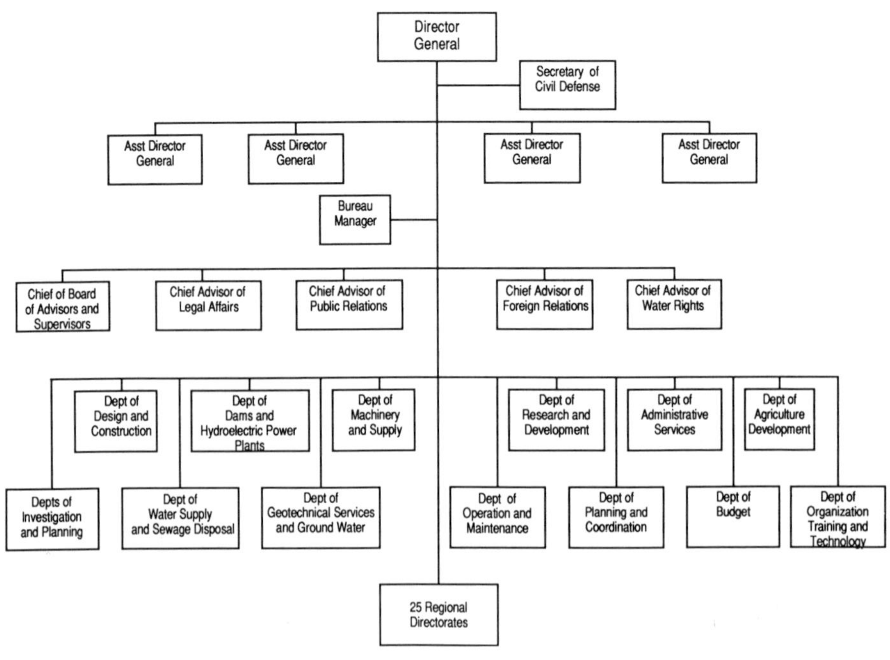

Figure 1.1. General Directorate of State Water Works (DSI). Source: DSI (1984b).

secondary or staff level is the Division office in Ankara. The tertiary level consists of the Field or Regional Directorate offices (Figures 1.2 and 1.3); there are 25 Regional Directorates dispersed throughout Turkey (Figure 1.4). Figure 1.5 is an example of line functions below the top staff division level. In 1983, DSI had a total of 25,702 employees.

The 25 Regional Directorates are divided into Central Regional Offices, Field Division Offices, and Field Section Offices. Major functional offices are: Mapping; Hydrometric Measurements; Land Management, Classification, and Drainage; Groundwater Activities; and Planning and Construction for Major and Minor Water Projects. These functions are interdependent but will be discussed separately. (See Table 1.2 and Map 11.2 for the population and location of these DSI regions.)

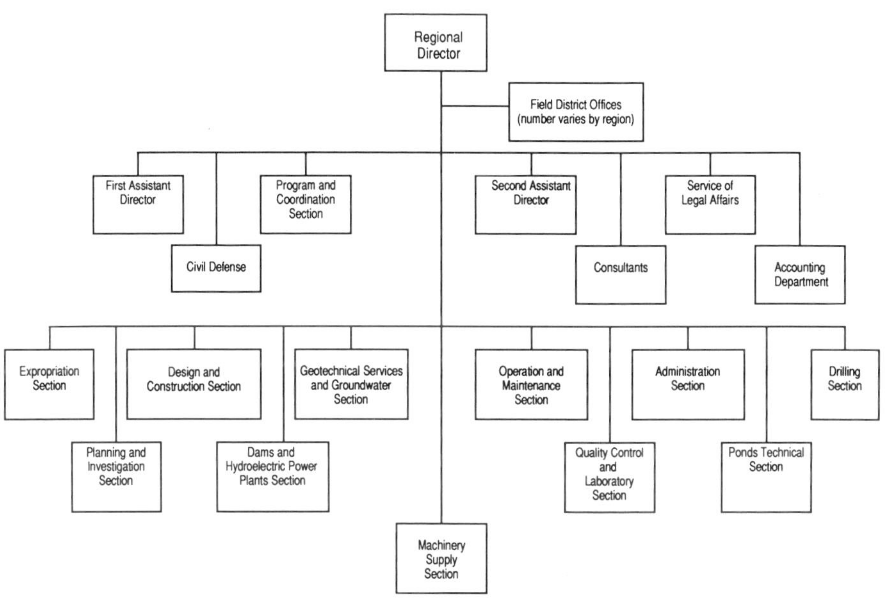

Figure 1.2. DSI regional directorate. Source: Based on discussions between DSI officials and Mitchell, Ankara, March 1985. See also DSI (1984b). Note: Representative of the 25 Regional Directorates in Turkey.

MAPPING   DSI's General Directorate of Mapping prepares maps for irrigation areas, dam sites, reservoir areas, and hydrographical areas. Standard scale is 1:5,000.

HYDROMETRIC MEASUREMENT   Accurate data are an essential first step in developing water resources. Hydroelectric projects, irrigation programs, and flood control programs are dependent on measurements of stream flow. DSI has established a hydrometrical network in its water basins. The Electrical Investigations Administration participates in data collection and provides stream flow gauging data to the national hydrometric network.

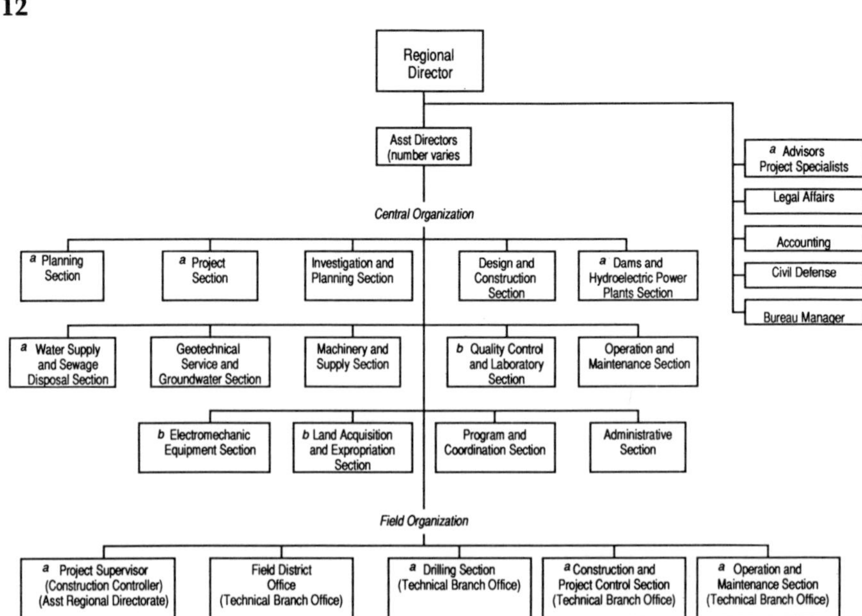

[a]Temporary positions (activated when required).
[b]Chief Engineers units which can be converted into Technical Sections or abrogated when not needed.

Figure 1.3.   Organization of DSI regional directorates. Source: DSI (1984b).

Stream flow gauging stations were first established in 1935. Data collection has expanded from larger basins and main tributaries to smaller streams. By 1983, DSI operated 1,144 gauging stations disbursed on the average of 1 per 678 sq km (DSI 1984b, 17).

Land Management, Classification, and Drainage   Since 1953, DSI has included entire drainage basins in its plans for water and land management. Initially, only the plains of Turkey had been included in the evaluation of soils, drainage, and irrigation potential. DSI conducts a four-stage program for water and land development: reconnaissance, initial land classification,

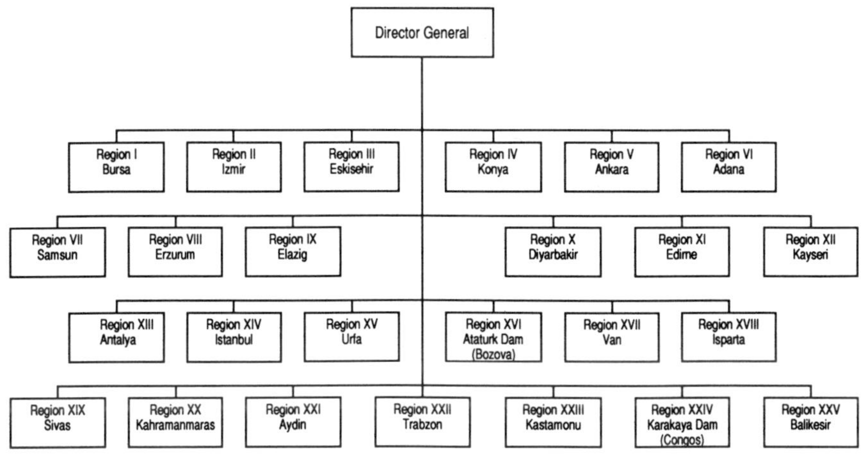

Figure 1.4. DSI regions in Turkey. Source: DSI (1984b).

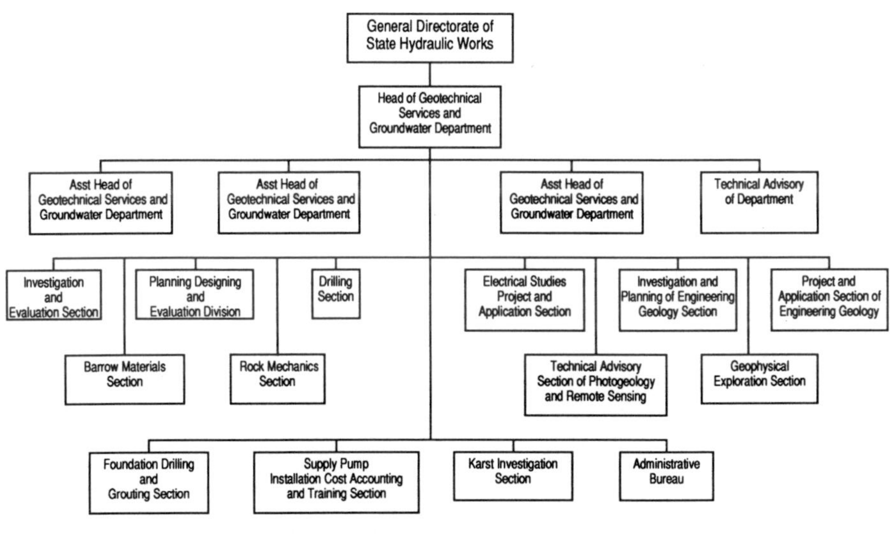

Figure 1.5 Line functions within DSI. Source: DSI (n.d.c.)

Table 1.2. Population of Turkey by DSI Region, 1960–1980

Region	1960	1970	1980
I	991,357	1,231,470	1,745,391
II	1,905,327	2,445,638	3,165,928
III	1,244,271	1,494,742	1,736,639
IV	1,238,836	1,607,376	1,983,050
V	2,362,225	3,224,679	4,156,707
VI	1,646,535	2,217,384	3,185,945
VII	2,068,582	2,543,585	2,963,685
VIII	1,554,035	1,898,525	2,139,929
IX	960,488	1,235,650	1,446,629
X	987,538	1,354,984	1,788,600
XI	792,431	875,510	1,007,436
XII	1,245,934	1,507,585	1,780,246
XIII	416,130	577,334	748,706
XIV	1,882,092	3,020,917	4,741,890
XV	401,919	538,131	602,736
XVI	575,404	847,798	1,184,423
XVII	947,484	1,133,230	1,273,801
XVIII	669,922	731,403	750,144
XIX	1,058,153	1,439,770	1,914,324
XX	1,192,401	1,443,673	1,693,971
XXI	1,602,798	1,934,606	2,076,574
XXII	1,002,679	1,190,255	1,405,458
XXIII	100,879	1,111,031	1,244,745
TOTALS	27,754,920	35,605,176	44,736,957

Source: DSI (1984b).

preliminary drainage studies, and detailed land classification and drainage analysis.

Reconnaissance studies survey agricultural land availability and present land use, identify areas with water problems, and prepare maps at 1:100,000. These studies have been completed for all of Turkey, with surveys, maps, and reports for a total of 49,995,905 ha (DSI 1984b, 17).

From 1954 through 1982, DSI had classified 10,116,558 ha

of land (DSI 1984b, 17). There are six land classes, of which classes I through IV are irrigable with increasing degrees of difficulty and cost of preparation. Class V land is temporarily deemed non-irrigable pending detailed technical and economic feasibility studies. Class VI lands are those that are impracticable for agriculture because of inaccessibility, slope, soil deficiencies, or urban use.

Drainage studies are essential for irrigation projects. Artificial and natural drainage are key factors in developing the optimal water table and controlling salinity. These studies are conducted as the third stage of water and land development potential.

After the previous three stages, a more detailed land classification and drainage analysis is undertaken. Areas are mapped at 1:5,000. These detailed studies take between two and three years to complete. Since 1970, DSI has suffered diminished capacity for detailed studies. It is trying to correct this insufficiency, but the lack of sufficient trained staff and technicians remains a management problem for DSI.

GROUNDWATER   Since 1959 DSI has conducted an active program of hydrogeological and geophysical investigations and exploration drilling during its reconnaissance phase to determine the location, amount, and quality of groundwater (see Figure 1.5). The final outcome of the reconnaissance phase is a "Groundwater Reserve Report" for the individual basin region that is under study. The report provides the data from which the planning phase begins. This phase analyzes groundwater reserves and allows planning for water development projects. In early 1983, analysis revealed an annual groundwater reserve of 9.5 billion m^3 which could irrigate 600,000 ha (DSI 1984b, ch. 4 and 5).

Planning and Construction for Major and Minor Water Projects
DSI categorizes its projects into major and minor works. About 5 percent of the annual budget is devoted to minor projects which affect small areas and require small investment. These provide benefits for small areas and relatively few people, but they offer a

quick economic and social return that can be used to lessen socio-economic problems. Major works benefit the national economy and include entire water basins.[3] All of Turkey's water basins have been investigated, and practically all planning for water resource development has been completed. Among major works, there are 194 projects in the planning stage.

Minor projects include new and repairable small gravity irrigation and pumping projects as well as small flood control and drainage projects. Surface and groundwater irrigation projects are usually developed by DSI and assumed by soil and water cooperatives for initial cost as well as long-term maintenance. DSI had drilled 5,712 wells and constructed 111 small detention dams by 1983. Additional minor projects included 2,262 flood control facilities to protect 192,375 ha and 1,862 settlements (DSI 1984b, 27).

DSI also maintains an operation and maintenance function as well as research and development.

Strategy for the Future

By 1985, Turkey had constructed and put into operation 100 dams. These multiple purpose dams served for irrigation (Table 1.3), flood control, hydroelectric energy, and for domestic and industrial water supply. Turkey plans eventually to construct 500 more dams, 430 of which will be hydroelectric power plants. DSI studies indicate that these goals will provide maximum utility of Turkey's potential water resources. There are 66 new dams and hydroelectric power plants already being constructed. The final design has been completed for an additional 30, and 31 more dams and hydroelectric power plants are in final design.

As has been pointed out (Kolars 1986b), Turkey is in a position to achieve agricultural self-sufficiency and possibly to have surpluses to sell to the rest of the Middle East. However, the economic

---

3. The term "water basin" is used here is place of "river basin." Several areas of internal drainage do not have a single river to serve as a focus of definition.

Table 1.3. Increase in Irrigated Areas in Turkey, 1963–1980

Development Plan	Year	Irrigated Area (in ha)
I	1963	268,856
I	1964	296,698
I	1965	332,329
I	1966	356,582
I	1967	447,364
II	1968	499,338
II	1969	558,993
II	1970	597,717
II	1971	619,998
II	1972	666,412
III	1973	715,901
III	1974	776,600
III	1975	838,015
III	1976	886,631
III	1977	933,704
IV	1978	968,086
IV	1979	1,016,640
IV	1980	1,050,462

Source: DSI (1984b).

future of southeastern Turkey and the future supply of energy to Turkey's national economy depend on the success of GAP.

GAP is an ambitious effort to reduce Turkey's energy deficit, raise the standard of living in eastern Turkey, and ease the political instability in an area of insurgency. As will be seen in subsequent chapters, GAP is much more than dams and irrigation canals—it is an integrated megahydroelectric and agricultural project that seeks viable development for Southeast Anatolia. GAP is a significant element of Turkey's national water and energy development programs. GAP's potential for success is great, as is its capacity to create a panoply of new problems.

2

The Southeast Anatolia Project:
An Overview

The Tigris-Euphrates is among the world's great river systems and Turkey's projects for developing and harnessing the potential of that watershed merit—for sheer size and complexity—a very honorable place among the hydroelectric development schemes of the modern world. The Southeast Anatolia Development Project (GAP) and related efforts are an integrated development plan encompassing a wide array of physical, social, and economic infrastructures. It is a massive undertaking of international significance, whose closest conceptual analogues are the American Tennessee Valley Authority or the Mekong Valley Scheme.

The area covered by this project is remote and the terrain is difficult. For several decades this watershed did not figure at the top of the list of the Turkish Republic's development priorities because of its relatively sparse population. Only when major needs of more populous regions had been addressed did Turkey turn its eyes to the Euphrates in the early 1960s.

For reasons both political and financial, Turkey has chosen as much as possible to build its Tigris-Euphrates projects largely on its own, using Turkish money, Turkish companies, and Turkish know-how—an admirable accomplishment for a developing nation. Due in part to this decision, Turkey is now an exporter of engineering expertise and a competitive bidder for development projects throughout the region.

A project of such magnitude inevitably carries with it major impacts. Turkey is not alone in its use of the Tigris and Euphrates river basins. The changes in the quantity and quality of the water flowing downstream to Syria and Iraq will be analyzed in depth in subsequent chapters. This is not, however, a totally negative projection, for Turkey alone, among all the nations at the eastern end of the Mediterranean Sea, has a sufficient surplus of fresh water which it may be willing and able to share with its parched neighbors.

Overview of GAP

GAP is a massive, planned development program within the Turkish portions of the Euphrates and Tigris river basins. The project includes land along the border with Syria and on the intervening plains between the two rivers (Map 2.1). The GAP will include dams, hydroelectric power plants, irrigation projects, and infrastructure supporting not only agriculture but other economic and social quality-of-life improvements, such as transportation, non-farm employment opportunities, and improved education and health services. Given that GAP will create economic, social, and spatial changes once energy and irrigation schemes come on line, the Turkish government views the project as a comprehensive "integrated regional development project."

Six provinces (vilayets or *il*) are in the project area: Adiyaman, Diyarbakir, Gaziantep, Mardin, Sanliurfa (frequently referred to as Urfa), and Siirt (Map 2.2). For organizational convenience, the GAP is divided into the Euphrates and Tigris development plans, of which the Euphrates portion is well under way and the Tigris portion is in the beginning stages. There are 13 large sub-projects altogether, 7 of which are on the Euphrates River (Lower Euphrates, Karakaya, Euphrates Border, Suruc-Baziki, Adiyaman-Kahta, Gaziantep-Araban, Gaziantep) and 6 on the Tigris (Tigris-Kralkizi, Batman, Batman-Silvan, Garzan, Ilisu, Cizre—see Map 2.3). These

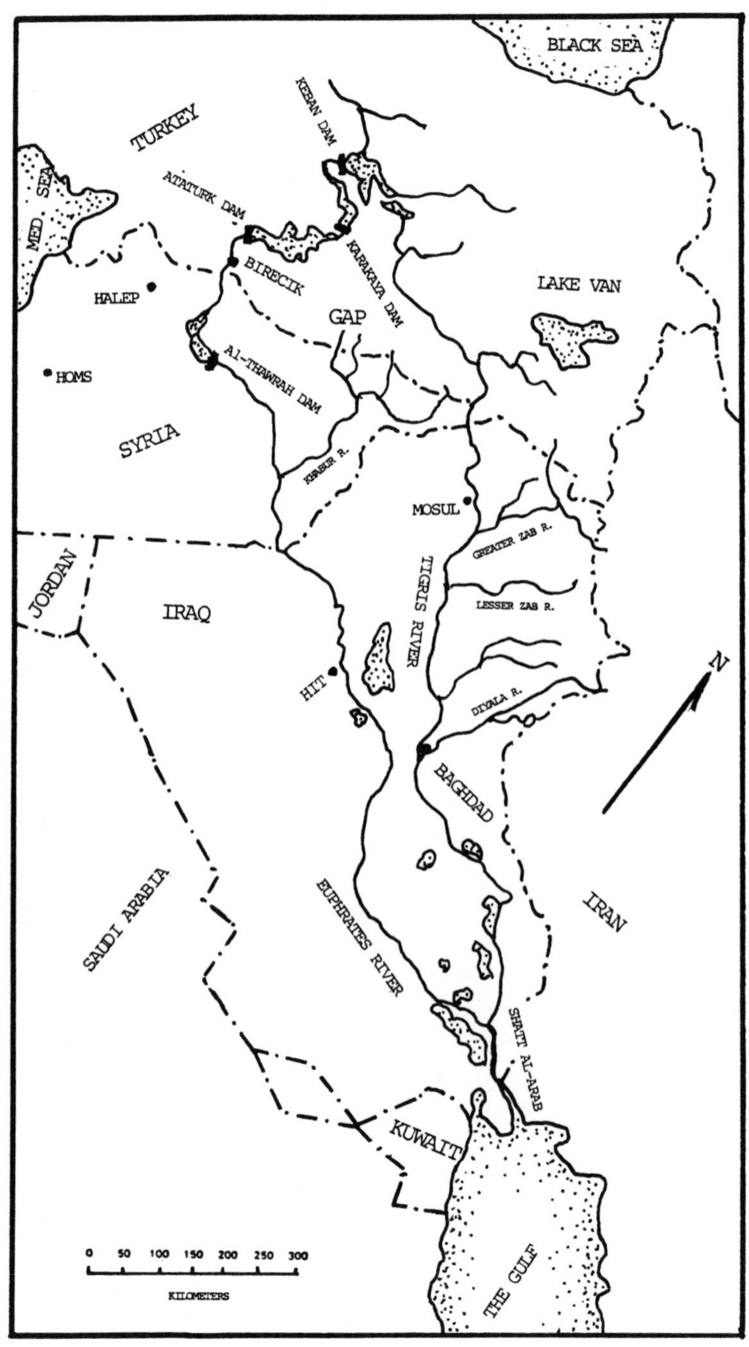

Map 2.1.   The Tigris and Euphrates river system.

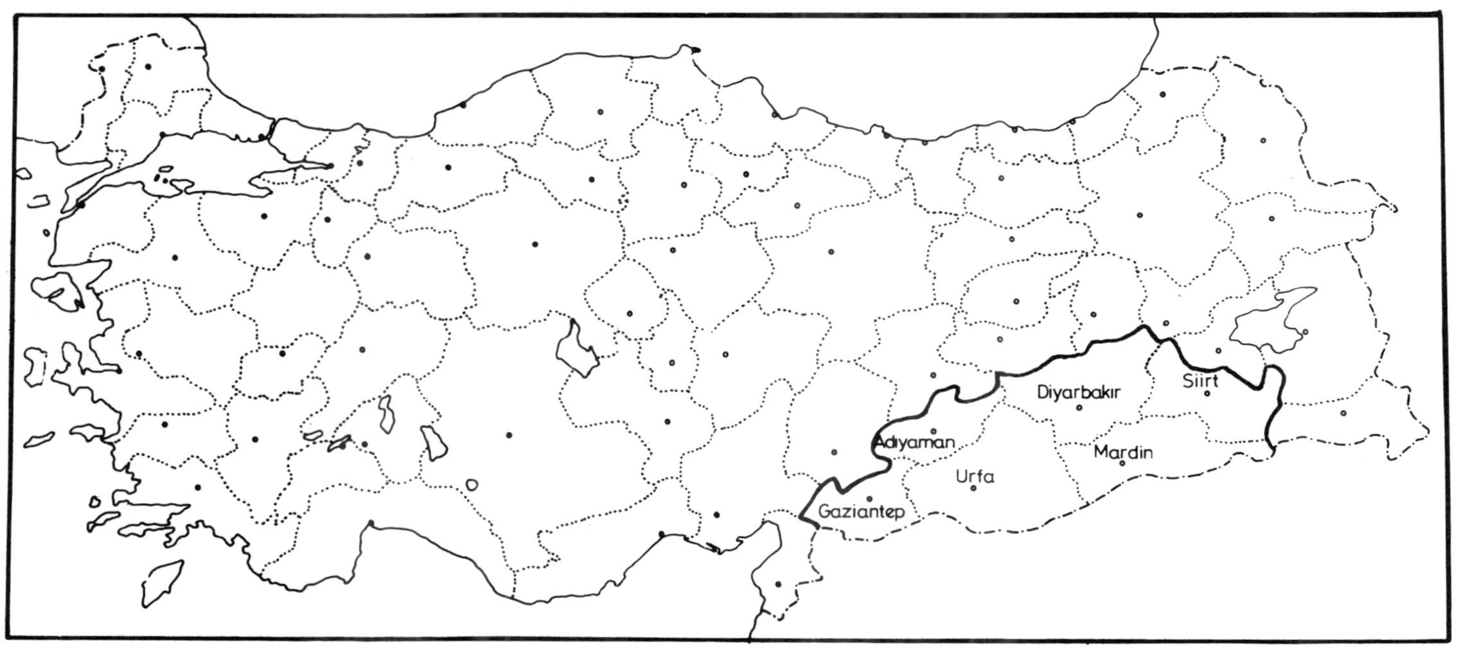

Map 2.2. Six provinces directly affected by GAP.

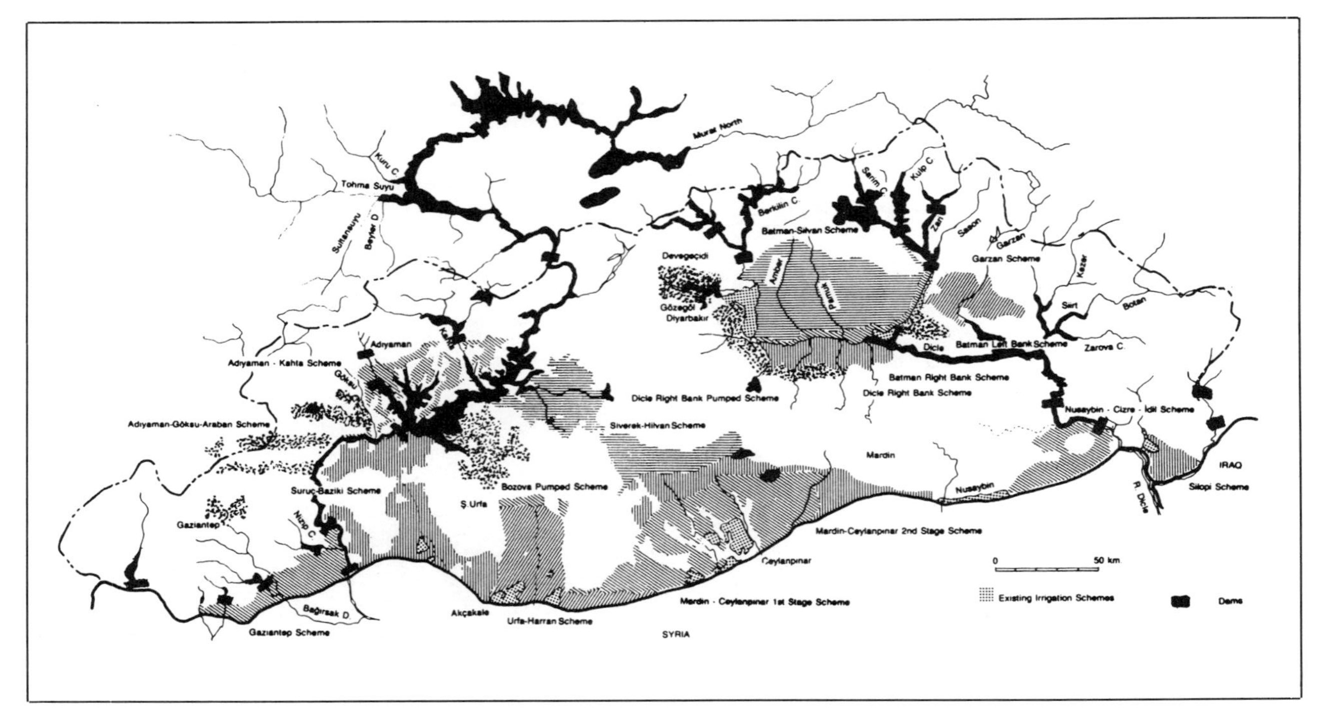

Map 2.3. The Southeast Anatolia Project (GAP).

Table 2.1. Benefits Expected After River Development in Turkey[a]

	Irrigated Area (in ha)	Flood Control (in ha)	Installed Capacity (in MW)	Energy Production (in GWh)	Water (in Mm³)
Euphrates	1,506,867	1,220	8,752.36	35,119	82.5
Tigris	525,336	—	3,405.68	12,644	—
TOTAL	2,032,203	1,220	12,158.04	47,763	82.5
ALL TURKEY	5,925,032	512,320	30,911.50	109,684	2,520.1

Source: DSI (1984b, 24a).

[a] Note that numerous estimates of development benefits have been published for GAP. Figures given in this table do not necessarily reflect those selected for analysis in the accompanying text. Moreover, the value given in this table for the Euphrates would include 1,280 kW installed capacity for the Keban Dam, which, technically, is not included in GAP statistics.

in turn include 15 dams, 14 hydroelectric power stations, and 19 irrigation projects. On the Euphrates, the Keban Dam is already completed and producing electricity, as is the Karakaya Dam on the same river. The latter's reservoir is filled, and the first of its electricity was produced in October 1987. Work on the largest dam of all, the Ataturk, is proceeding at an accelerated pace (*MEED* 1985a). Although the Keban Dam, upstream on the Euphrates near Elazig, is not considered part of GAP, its presence makes it an integral element in the management of the river.

Turkey's economic planners are optimistic about GAP (Table 2.1). They believe the complex of hydroelectric dams and irrigation canals will greatly improve the economic and social life of this area, which in the past has lagged behind that of the rest of the nation. According to them, in less than 30 years, when GAP is scheduled for completion, about 2 million ha of land will be irrigated in the Euphrates-Tigris river basins, including the GAP project area. This would be a significant increase representing 64 percent of the total public and private lands presently irrigated in Turkey (3,176,330 ha in 1983) (DSI 1984b, Table 4.1). The change will be dramatic, for in 1986 only 4 percent of Turkey's irrigated fields were located in the six GAP provinces.

Table 2.2. Installed Electrical Capacity and Production in Turkey and Selected Countries, 1981
(GAP capacity shown for comparison)

Country	Installed Capacity (in MW)	Production (in GWh/ yr)	Production as Percent of Potential Production
Turkey	2,171	12,616	11.5
GAP (potential)	(7,620)	(22,000)	
France	19,500	70,682[a]	70.7
Italy	15,766	45,736	80.2
Yugoslavia	5,510[b]	26,558	40.2
Greece	1,714	3,408	16.5
Portugal	5,616	5,193	30.6
Turkey in 1984:			
Hydroelectric	3,875	13,423	
Thermal	4,584	17,185	
Total Turkey in 1984	8,459	30,608	

Source: DSI (1984b, 245–246).

[a] Figures for 1980.
[b] Figures for 1979.

GAP eventually will increase Turkey's installed hydroelectric capacity by 7,620 MW, double the present amount, and is expected to generate 22 billion kWh, or nearly two-thirds more than the amount of electricity produced hydroelectrically in the entire country (Table 2.2) in 1984.

GAP is an expensive endeavor, costing in 1981 prices about $30.26 billion (TL 2,270 billion). However, using the 1981 price index, GAP when finished will contribute annually about $933 million or TL 70 billion to the Turkish economy (in 1981, $1 = TL 75; by 1990, $1 = TL 2,700). Preliminary estimates are by definition tentative; consequently, adjustments to costs must be made throughout the development cycle of this project.

The entire project for both the Tigris and Euphrates river basins is scheduled for completion by the year 2013. This completion date is also subject to revision and redefinition; the progress of GAP and its future prospects are discussed in detail in the final chapter of this work.

The Development of GAP: A Chronological Review

The Euphrates and Tigris rivers attracted the attention of planners in the early years of the Turkish Republic, but the remoteness of the region and the more pressing matters facing the young nation prevented action at that time. A gauging station was established at Kemaliye near the present site of the Keban Dam in 1936. This was among the first such in Turkey. Development planning for the region was further delayed by World War II. The General Directorate of State Water Works (Devlet Su Isleri/DSI) was established in 1953. This expedited large-scale hydroelectric planning, in which the Keban Dam and GAP were to assume an important role.

The development of large hydroelectric projects was favored by the fact that Suleyman Demirel, former Premier and former chairman of the disbanded Justice Party, received his degree in hydrological engineering from the Technical University of Istanbul and continued his education through an Eisenhower Exchange Fellowship in the United States. Upon his return to Turkey in 1955, he became, at the age of 31, the Director of DSI, where he earned the nickname, "King of Dams." When his fortunes led him into politics he became Prime Minister following the Justice Party victory of 1965. Following that election he helped obtain $300 million for construction of the Keban Dam and on 18 October 1976 presided at the laying of the foundation for the Karakaya Dam. It was during this period that the plans for the GAP were prepared and the search for credits for the Ataturk Dam was begun (*TDN* 1986).

During this same period, Turgut Ozal, destined to become Prime Minister in the 1980s, finished graduate work in the United States in economics and engineering and returned to Turkey to serve as General Deputy Director of the Electrical Studies and Research Administration in Ankara during the late 1950s. At that time he directed studies of the Euphrates and Tigris rivers and of their hydroelectric potential. A series of reports which emphasized the potential of the region were authored and subsequently published by Korkut Ozal, younger brother of Turgut, which emphasized the potential of the region (Ozal et al. 1967). In the political campaigns of the mid-eighties, both Turgut Ozal and Suleyman Demirel would claim credit for the Keban Dam and the inception of GAP. Certainly both Ozal and Demirel were important in the early years of these projects.

In any event, the potential of the Euphrates region was recognized and projects for the development of its soil and water resources were begun in 1961. A "Euphrates Survey Report" and another "Lower Euphrates Project Survey Report" were produced in 1964 and 1966, respectively.[1]

In the meantime, preliminary work on the Keban Dam farther upstream began in 1964–1965, the same year that the Ministry of Power and Natural Resources was established. The Keban Dam project was seen as part of a much larger scheme, although to this day it is not included within GAP. In 1968, the Keban Dam was reported as being able to provide "irrigation of 1.65 million ha and 600 million kWh electricity by 1972" (*Turkish Economic News Summary*, No. 310, 17 June 1966, cited in Hershlag [1968, 193, note 4]). Since the Keban Dam is solely a hydroelectric project not intended to provide extensive irrigation, and since the figure of 1.65 million ha corresponds reasonably to amounts predicted for GAP downstream, it seems safe to assume that the development of the entire Euphrates was being discussed at the time.[2]

The main contract for construction of the Keban Dam was signed 19 February 1966, between the Turkish government and SCI-Impreglio, a Franco-Italian consortium. Work was financed

1. This author has not examined either of these reports. They are referenced in Sonmez et al. (1985, 5).

2. Although the Keban Dam has consistently been referred to in Turkish publications as solely a hydroelectric generating facility, a recent news release (*Newspot* 1987, 5) refers to a proposed "Kuzova Project" by means of which "300,000 donums" (30,000 ha) of land would be irrigated by waters from the Keban reservoir. Computations in this analysis do not include any water depletions that might result from such a project were it to be fully implemented.

by the European Investment Bank, USAID, and the French, German, and Italian governments (Tansey 1982; *Turkish Economic News Summary*, No. 295, 25 February 1966, cited in Hershlag [1968, 223, note 5]). Total cost in 1974 dollars was about $85 million (TL 8,000 million) (Graham-Brown and Barchard 1981). The foundation of the dam was laid 12 June 1966, and its reservoir was filled and power production begun in 1974.

At that time, the project was described as a connecting link helping to integrate the separated power sub-grids of the country as well as providing much needed energy for both the northwest and the southeast regions (DSI n.d.a). At the same time it was written, "Planning of the Euphrates catchment area is closely concerned with the construction of the Keban Dam where 4 eastern provinces, 30 counties, and 1820 villages are concerned" (TMRR n.d.). The Second Five-Year Plan, 1968–1972, included construction of at least four dams in the southeast: Keban (1974), Surgu (1969), Medik (1975), and Cip (1965).

Work continued apace. By 1970, electrical production plants and transmission systems owned by the Iller Bank and DSI were transferred to the Turkish Electrical Authority (TEK) by law number 1312 (TSIS 1988, 232). In 1968, two years earlier, contract bids were solicited from foreign and domestic firms for feasibility studies for additional hydroelectric stations and for a master irrigation plan, both of these items having received priority in the survey reports mentioned above. These studies were completed in 1970. A series of works were suggested, to be carried out in the following order:

1. Karakaya Dam
2. Golkoy Dam
3. Middle Karababa Dam
4. Bedir Pumping Station
5. Bedir Dam
6. Urfa Tunnel
7. Hilvan Pumping System.

The reports indicated that 700,000 ha would be irrigated as a result and that 2,700 MW of installed capacity would produce 14.8 billion kWh/yr. However, the subsequent leap in world petroleum prices required a revised analysis of energy and pumping projects in Turkey. As a result, a new report, the "High Karababa Dam and HES [*sic:*, Hydroelectric Station] Summary Report" recommended that the Middle Karababa Dam be raised 60 m and that the Golkoy and Bedir Dams were unnecessary. It also meant that the Bedir Pumping Station was no longer needed and that energy consumed by the Hilvan Pumping System would be reduced a significant amount. The "Urfa Hydroelectric Central and Harran Ova Irrigation Planning Report,"[3] which was prepared in the same year, separated 142,000 ha of irrigated land south of Harran from the Mardin-Ceylanpinar irrigation scheme and gave priority to the former.

Despite difficulty in obtaining full outside funding, contract bidding for work on the Karakaya Dam was called in 1976. The reservoir behind the dam began to fill in November of 1986, and the third generating unit began power production in October 1987. The cost to date is estimated at about $900 million (*MEED* 1987b).

Also in 1976, the 12 sub-projects that now constitute GAP were combined under that title (TMENR 1980). Again, it should be noted that although the Keban Dam is an integral part of the overall Euphrates development scheme, it does not fall within GAP.

Contracts for a high dam at the Middle Karababa site, subsequently renamed the Ataturk Dam and reservoir, were let in 1983. That project is scheduled to begin electrical production in the early 1990s.

In 1987 GAP included an area of 74,000 sq km lying between the Anti-Taurus Mountains and the Syrian border. At that time approximately 42,000 ha were irrigated with surface and underground waters in both the Euphrates and Tigris basins (Table 2.3). The Keban Dam, not technically part of GAP, was producing hydroelectric power. The Karakaya Dam was nearly completed and already producing some electricity. In the eastern part of GAP, the Cagcag and Botan hydroelectric stations had a combined

---

3. These reports were unavailable to the author. They are referenced in Sonmez et al. (1985, 5).

Table 2.3. Existing Irrigation from Ponds and Undammed Waters

Name	District	Date Open	Purpose	Height (in m)	Capacity in Mm^3	Area Irrigated (in ha)
Gozegol	D.Bakir-Merkez	1964	I	8.75	8.0	550
Kurtkayasi	D.Bakir-Merkez	1970	I	5.00	0.1	8
Ortaviran	D.Bakir-Cinar	1963	I,A,D	8.45	2.0	516
Desan	D.Bakir-Mazidagi	1961	A,D	4.00	0.125	—
Dirsekli	Mardin-Idil	1968	I,A,D	14.00	1.62	183
Serifbaba	Mardin-Derik	1973	I,A,D	14.70	1.60	115
Kunres	D.Bakir-Cinar	1974	I	6.50	0.594	18
Bespinar	D.Bakir-Cinar	1976	I	8.00	1.0	200
Yildiztepe	Mardin-Derik	1975	I		1.5	221
Halilan	D.Bakir-Cermik	1978	I	21.00	—	550
Hacikamil	Urfa	1966	I	—	—	450
Batman	Diyarbakir		I	—	—	3,500
Nusaybin	Mardin		I	—	—	5,750
Akcakale	Urfa	1977	I	—	—	13,800
Ceylanpinar	Urfa	1979	I	—	—	9,000
Devegecidi Dam	Diyarbakir	1972	I		195.04	7,500
					TOTAL	=42,361

Source: Sonmez et al. (1988, Tables 1 and 2).

Purpose: I = irrigation
        A = animals
        D = domestic

operating capacity of 16 MW. Compared with these modest achievements, the projected return on the completed GAP is impressive. Table 2.4 shows the anticipated results of such development in the year 2010. The figures presented in this table must be taken with some caution as they tend to reflect the optimism of planning groups closely associated with the project. Nevertheless, the six provinces found within the GAP will provide nearly one-fifth of all Turkey's hydroelectric energy, irrigated lands, and surface waters in the years ahead.

Table 2.4. Projected Potential—Southeast Anatolia Region
(The 6 GAP provinces estimated for 2010)

Item	Region	Turkey	Region as Percent of Turkey
Population (000)	7,572	82,234	9.2
Workforce (000)	4,590	50,234	9.1
Total Area (sq km)[a]	72,958	774,815	9.4
Projected Irrigated Area (000 ha)[b]	1,800	8,500	21.2
Surface Waters (Mm^3/year)	47,000[c]	180,000	26.1
Hydroelectric Energy (billion kWh/yr)	22	110	20.0
Phosphate Reserves (million tons)	453.8	465.6	97.5
Asphalt Reserves (million tons)	39.5	39.5	100.0

Source: Sonmez et al. (1988, 33, Table 18).

[a] Excluding lakes.
[b] GAP area.
[c] Euphrates and Tigris basins.

Financing and Associated Problems

As with all vast development projects—as has been, for example, the continuing case with the development of the Colorado River—there is concern both nationally and internationally about the feasibility, practicability, and efficiency of GAP.

Syria and Iraq are obviously interested parties, concerned about construction activities that could reduce the annual flow of the Euphrates and Tigris through their countries (*Middle East Markets* 1984). Both nations have important programs including dams and irrigation schemes on the rivers and are reluctant to see a sizeable portion of the annual flow removed for irrigation in Turkey (Graham-Brown and Barchard 1981). The World Bank and most international funding agencies are not prepared to support water projects involving international riparian rights unless all affected countries agree upon use of the water involved. Turkey realizes this and argues that the GAP will benefit everyone (that is, Syrians and Iraqis as well as Turks). The Director of Turkey's DSI has stated: "The annual flow levels of the Euphrates vary from 162 to 700 m^3/s, depending on the season. Once the GAP is completed, there should be a steady flow of 500-600 m^3/s" (Graham-Brown and Barchard 1981).

In July 1984, Iraq agreed with Turkey to accept a minimum flow of 500 m^3/s, but Syria refused to negotiate at that time (*MEED* 1984b; *TDN* 1984a). Discussions between Syrian and Turkish leaders began in 1986 and are continuing. In July 1987, Premier Turgut Ozal visited Damascus; upon his return he announced the signing of a protocol with Syria guaranteeing 500 m^3/s of Euphrates flow across the border below Birecik into Syria. This agreement does not address the July 1984 agreement with Iraq for a similar amount of water.[4]

Turkish officials originally requested $1.5 billion for the Ataturk Dam, $150 million for tunnel projects associated with it, and $200 million for equipment from various international sources (*Middle East Markets* 1984). Because of Syrian and Iraqi reservations about Turkey controlling the Euphrates' flow, the World Bank hesitated in providing hard currency. Syrian and Iraqi objections about GAP also influenced other international lending institutions to refuse support (*Middle East* 1984). Consequently, tenders for the civil, electrical, and mechanical contracts were delayed. For example, although the Karakaya Dam's civil works contracts were let to an Italian firm in September 1976, Turkey could not produce the necessary foreign currency because of serious exchange shortages (*MEED* 1981a). As a result, contracts are

---

4. The physical situation is complicated by the configuration of the Turkish canal and tunnel system which will bring large quantities of water to the headwaters of the Khabour and Balikh rivers in the Syrian Jezirah. Not only will the flow of those streams be altered but the quality of their waters will be affected as well. Such considerations are typical of the potential for conflict or cooperation wherever Middle Eastern rivers are concerned. This situation has been analyzed and commented on by many, see, for example, Naff and Matson (1984). The potential areas of disagreement mentioned above as well as other similar situations relating to GAP will be examined in the chapters that follow.

now being awarded to Turkish firms, and international companies such as Bechtel appear to have lost out (Barchard et al. 1983).

Regardless of financing difficulties, GAP officials have proceeded with development and construction. In 1983, a contract to build the Ataturk Dam was awarded to the Turkish companies of Palet Insaat, Seri Insaat, and Energy-Su—renamed the ATA-Insaat Consortium. The award of the Ataturk Dam project to Turkish contractors evoked initial skepticism in both local and international construction circles, but ATA-Insaat has proven it does have the resources and technical competence to complete the project (*MEED* 1984b). For example, exclusively Turkish contractors built the 179 m high Hasan Urgulu Dam in Samsun Province between 1972 and 1979.

Such positive attitudes subsequently paid off. In March 1985, the Export-Import Bank of New York and the Manufacturers Hanover Trust loaned Turkey $111 million for the Ataturk Dam (BBC 1985), and European banks are now providing $460 million for equipment purchases. As a result, the first of eight 300 MW generators is scheduled to start operation at the Ataturk station in May 1991 (*MEED* 1985b). The seven other generators are expected to be on line by 1994. Meanwhile, Swiss banks and the World Bank, albeit reluctantly, loaned funds for the Karakaya Dam which, originally scheduled for completion in 1989, had its first generator come into production in the spring of 1987.

The government of Turkey is considering raising a bond issue for the Ataturk Dam, as it did with the Ataturk Bridge across the Bosporus (*MEED* 1981b). Prime Minister Turgut Ozal, architect of Turkey's economic recovery since the reconstitution of the government by the military in 1980, believes that the GAP will "produce enough food, clothes, and other articles for an additional 20 million people" (*Euromoney* 1984). Ozal goes on to state that electricity from the Ataturk Dam project will pay for itself and for the associated irrigation tunnels in four years. His projection is based on selling electric power at 1984 prices and does not include increased GNP through increases in agricultural production. Further elements of this development are indicated in Table 2.4.

There are additional problems beyond financing. The dam

Table 2.5. Average Sizes of Selected Settlements in GAP Region Provinces and Counties

Province and Settlement	Average Village Size Selected County	Average Village Size in Province
Adiyaman		658
Samsat	405	
Kahta	789	
Diyarbakir		526
Bismil	433	
Cermik	435	
Cinar	517	
Cungus	378	
Elazig		358
"57 villages with 70,000 population"		1,228
"232 villages with 64,000 population"		276
Average settlement size in Turkey		650

Source: TSIS (n.d.).

sites are in remote, mountainous areas, difficult of access. Infrastructure is lacking and roads, workers' accommodations, and ancillary services must be provided. As a result, even with international financing, the Keban Dam was completed four years late. There also remains concern about shortages of skilled workers, from laborers to engineers, which may affect GAP's completion (Tansey 1982).

The reservoirs impounded behind the major GAP dams have necessitated and will in the future require the resettlement of large numbers of villagers from flooded areas. Estimates of dislocated settlements vary. Some sources have estimated 70,000 people from 57 villages must be relocated from the reservoir areas of the Karakaya and Ataturk Dams (*TDN* 1987; Parry 1984). A Turkish television report in December 1985 mentioned that 94 villages would be moved for the Karakaya Reservoir and 138 for the Ataturk Dam, with a total of 64,000 people from the 232 villages. This would be an average of 276 persons per village, an unusually small population for any given settlement. Table 2.5 gives some average settlement sizes for selected political units in the GAP region. The counties (*Ilce*) chosen are among those bordering the Euphrates River. It is difficult to pursue this question without knowing the exact names of the settlements involved. The 232 settlements may refer to 232 *mahalle* (plural *mahalleler*) or neighborhoods within villages. These are sometimes discrete settlement units, particularly in rough terrain. What does seem reasonably consistent is the 64,000–70,000 population involved.

This problem may well be further exacerbated by the fact that many of the displaced villagers are Kurdish. This is corroborated by the census books, which give both Turkish and Kurdish names for many of the settlements in this region. The entire Kurdish question is one that raises issues of security and regional economic development in a critical context. While it is not the purpose of this book to engage in political speculation, some further mention of the subject will be found in the concluding chapter.

Despite such hurdles, the Southeast Anatolia Project is moving steadily toward completion. GAP in its entirety promises to reshape southeastern Turkey—and perhaps all of the Tigris-Euphrates drainage basin beyond Turkey's borders.

The Euphrates Portion of GAP

The Euphrates project will eventually have a combined hydroelectric capacity of 5,440 MW and a potential to irrigate 1.5 million ha (3.7065 million acres).[5] GAP has, in its scope, been compared with projects such as the Tennessee Valley Authority. Given its many facets, including irrigation, it is one of the most comprehensive developments attempted anywhere (see Tables 2.3 and 3.1). The linchpins of this regional development are three major dams: the Keban Project, the Karakaya Project, and the Ataturk Project (*Newsweek* 1982). (See Table 2.6.)

---

5. Figures vary considerably from source to source. Introductory estimates in summary statements will be clarified in subsequent chapters.

The Keban Project
The Keban Dam is a compacted, rock-filled dam with a clay core and a total embankment volume of 15.6 Mm³. The dam's foundation height is 210 m, with a crest length of 1,125 m. Its reservoir holds 30,600 Mm³ and has an area of 675 sq km, although in 1982 it was reported as covering 680 sq km (*Newsweek* 1982). A major seepage occurred in 1974, but the cavern which caused it has subsequently been filled. The dam is intended to regulate the seasonal fluctuations of the river and to generate power (see note 2, p. 26).

In July 1981 the Keban Dam was producing electricity from four of its turbines. The fifth turbine was on line by December and the remainder came on line at three-month intervals starting March 1982 (Graham-Brown and Barchard 1981). The last four units, resulting in 1,280 MW final installed capacity, entered production in 1983 (*Modern Power Systems* 1985). During full production, the average annual electricity generated is expected to be approximately 6 billion kWh (TMENR and DSI 1985). Total production from 1974 through 1978 with fewer turbines was 17 billion kWh (DSI n.d.b).

With the completion of the Keban Dam and reservoir, the government announced that it had established river flow at a minimum of 450 m³/s and a maximum of 1,000 m³/s (DSI n.d.b). There is some evidence that significant variation in flow continues; moreover, reduced river flow during the filling of the reservoir resulted in a downstream crisis between Iraq and Syria (see Chapter 5). The impact of such irregularity on Turkish, Syrian, and Iraqi projects downstream will be considered in the chapters that follow.

The Karakaya Project
The Karakaya Dam is located about 166 km downstream from the Keban Dam, near Cungus in the Diyarbakir region. The Karakaya, with planned 1,800 MW capacity, was completed in late 1988. Total cost was estimated at $500 million in 1981 (Graham-Brown and Barchard 1981), toward which the World Bank committed $350 million (Tansey 1982). Annual electricity generation will be

Table 2.6. Technical Parameters for Major Dams on the Euphrates in Turkey

Characteristics	Ataturk Dam	Karakaya Dam	Keban Dam
Location	180 km downstream of Karakaya	166 km downstream of Keban	Elazig
Stream	Euphrates River	Euphrates River	Euphrates River
Purpose	Energy, irrigation, flood control	Power generation	Power generation
Date of Completion	Scheduled for 1989	Began filling in 1986	1974
Catchment Area	92,240 sq km		64,100 sq km
Annual Average Precipitation	750 mm (est.)	825 mm (est.)	925 mm (est.)
Annual Inflow	26,585 Mm^3	23,600 Mm^3	19,999 Mm^3
RESERVOIR			
Maximum Water Level	542 m	693 m	845 m
Total Reservoir Volume	49,000 Mm^3	9,540 Mm^3	30,600 Mm^3
Reservoir Surface Area	817 sq km	298 sq km	675 sq km
DAM EMBANKMENT			
Type	Rock, rock fill	Concrete, archgravity	Compacted rock fill with clay core
Height from Foundation	179 m	187 m	211 m
Height from Riverbed	169 m	137 m	167 m
Crest Length	1664 m	462 m	1125.72 m
Embankment Volume	86 Mm^3	2 Mm^3	15.5 Mm^3

SPILLWAY			
Type Gravity	Radial gate (dimensions 16 × 17 m)	Radial gate	Ogee type, concrete
Number	6	10	6
Discharge Capacity	24,000 m^3/s	17,000 m^3/s	12,000 m^3/s
DIVERSION TUNNELS			
Type	Horse-shoe	Circular, concrete lined	Horse-shoe section
Number of Tunnels	3	2	2
Tunnel Lengths	1326 m, 1376.20 m, 1396.40 m	568 m and 698 m	708 m each
POWER HOUSE			
Installations Turbine Type	Vertical Axis Francis	Vertical Axis Francis	Vertical Axis Francis
Number of Turbines	8	6	8
Discharge Capacity of Each Turbine	218.3 m^3/s	233 m^3/s	180 m^3/s
Installed Capacity	8 × 300 = 2400 MW	6 × 300 = 1800 MW	8 × 155 = 1240 MW
Average Annual Power Production	8,100 million kWh	7,500 million kWh	5,900 million kWh

Source: DSI (1981) and various other sources.

between 7.3 and 7.5 billion kWh (*Newsweek* 1982; TMENR and DSI 1985).

The Karakaya is a concrete arch-type dam with a height of 187 m and a crest length of 462 m. Its impoundment holds about 9500 Mm^3 of water (DSI 1982). Its embankment contains about 2 Mm^3 of concrete. Original plans were to attain power production by 1986, but this was delayed until 1987 (DSI n.d.b). The Karakaya Dam is similar to the Keban Dam in that its purposes are to regulate stream flow and generate hydroelectric power.

The Ataturk Dam, Power Station, and Irrigation Project

The Ataturk Dam site is located near the town of Bozova, 70 km northwest of Urfa and 181 km downstream from the Karakaya Dam. This project is one of the most ambitious engineering feats ever attempted. It will be the largest dam in Turkey; its filled reservoir capacity as well as its embankment volume will make it the fifth-largest dam in the world (see Tables 2.7 and 2.8) (*TDN* 1987; *Euromoney* 1984; Barchard et al. 1983).

The dam's 180 m high clay core and rock-filled wall will impound an 817 sq km lake with a volume of 48.7 Mm^3. The dam is built with rock excavated at the site, sand and gravel from the river bed, and basalt rock transported from a site about 4 km away. The volume of the Ataturk Dam's embankment will be 84.5 Mm^3, making it about five times larger than the Keban.

This third dam on the Euphrates will have a generating capacity of 2,400 MW from eight Francis turbines. Plans are for generation to be controlled by the TEK in Ankara. The reservoir will provide irrigation for about 730,000 ha, making it the first or second largest irrigation from a single source scheme in the world (*Modern Power Systems* 1985). The value of surplus food production available for export for the region to be irrigated by the Ataturk Project is estimated at $5 billion annually (*MEED* 1984b). This will be produced on land in the Harran, Siverek, Mardin, Ceylanpinar, and Hilvan plains. The irrigation system will be the largest in the world, containing a twin-bore, 24.6 km main tunnel, 283 km of main canals, 150 km of secondary canals, and 200 km or tertiary distribution canals.

Table 2.7. Additional Technical Parameters for Ataturk Dam and Power Plant

Average Flow	1026 m³/s
Highest Recorded Flow	6160 m³/s
Riverbed Elevation	380 m
Active Reservoir Volume	19,300 Mm³
Dam Peak Elevation	594 m
Diversion Capacity	2100 m³/s
Bottom Outlet	2 × 3 sliding gates
Bottom Outlet Capacity	1500 m³/s
Spillway	Controlled sill and canal
Auxiliary Gates	2 (5 m × 8.1 m) radial
Auxiliary Spillway Capacity	1000 m³/s
Intake Structure	Concrete gravity 8 entrances
Intake Elevation	593.6 m
Intake Gates	8 sliding (4.8 m × 7.7 m)
Penstocks	8 in the right abutment
Diameter of Penstock	Approximately 7 m
Length of Penstock	Approximately 600 m
Powerhouse	Closed type
Turbine Capacity	400,000 horsepower
Generators	Vertical axis
Generator Capacity	315 MVA
Frequency	50 Hz
Speed	150 rpm
Transformers	3 × 8 = 24 single phase
Switchyard	Open type

Source: DSI (1981).

General Kenan Evren, then Turkish Head of State, formally inaugurated work on the project in October 1981, the centennial of Ataturk's birth. The original plan was for completion by 1990 with the first turbine coming on line in 1987 (*Newsweek* 1982). Completion date for the project was delayed from 1990 to 1993 because of credit difficulties, but recent government decisions

Table 2.8. The World's Largest Dams

Name of Dam [Year Completed]	Location	Embankment Volume (in Mm³)	Reservoir Capacity (in Mm³)
New Cornelia Tailings [1973]	Arizona, USA	209.500	
Pati (Chapelton) [UC]	Argentina	200.000	
Tarbela [1976]	Pakistan	121.720	
Fort Peck [1940]	Montana, USA	96.049	
Ataturk [UC]	Turkey	80.500	
Yacyreta-Apipe [UC]	Paraguay/Argentina	81.000	
Daniel Johnson [1968]	Canada		141,852
Aswan High Dam [1971]	Egypt/Sudan		115,000[a]
Bennet WAC [1967]	Canada		70,309
Cabora Bassa [1974]	Mozambique		63,000
Ataturk [UC]	Turkey		48,700
Hoover [1936]	Arizona/Nevada, USA		35,154

Source: USDI (1987).

[a] The maximum capacity of Lake Nasser is 164,000 Mm³, but the maximum to which the reservoir has been filled to date was 115,000 Mm³ in October 1975 (Waterbury 1979, 116–53). The actual volume is significantly lower since the African drought.

restored the completion date to the end of 1990 (*MEED* 1985a). Consequently, the first turbine should be operating in 1991, with the remainder on line by 1993.

Cost of the entire project has been estimated at from $2 to $4 billion (*Newsweek* 1982; *Euromoney* 1984). The ATA-Insaat Consortium has made significant progress on the project. In June 1984, there were 800 workers stripping the site, building roads, and constructing a work camp with offices, stores, workshops, living areas, and schools for workers' children. Excavation has subsequently been completed on the power house and about 30 km of roads were built to expedite heavy equipment movement. Contracts for the dam's turbines, generators, and electrical equipment were placed in March 1984 (Table 2.7). Though its size varies depending on the season and the tasks at hand, the work force has increased to more than 3,000.

The most difficult part of the dam's construction was 1,200 km of drilling and grouting that had to be completed after diversion of the river was achieved. The grouting is in galleries driven under the dam's foundation and in the walls of the valley (*MEED* 1984b).

Geological Conditions at the Ataturk Site
The Turkish Surveying Administration began investigating possible sites for the Ataturk Project in 1964 (DSI n.d.b). The dam site that was chosen is located where the Euphrates enters a mountain pass. The dam is located in a wide valley at the beginning of the pass, rather than in the pass, because drilling and excavation revealed unsuitable geological instability farther down the canyon.

The rock at the site itself is limestone in various forms. The pass bottom is covered with rubble, worn rock waste, and river alluvium. Terrace deposits of sandy gravels 5 to 10 m thick cover the bedrock. Valley alluvium does not exceed 9 m in depth, and the dam body is set on a foundation of plicated limestone. The axis of the dam lies entirely within a thin bedded, slightly marly limestone (*Modern Power Systems* 1985).

The slopes of both abutments were originally covered by a talus mantle about 1 m thick. The foundation rock is dense, moderately hard, and homogeneous over the dam foundation zone. A main fault runs east to west about 1 km south of the dam but does not interfere with the dam. However, the dam site is only 240 km southeast of the East Anatolian fault zone (Mitchell 1977). Since the dam is in an active seismic zone, the dam design is said to be able to withstand an earthquake of a magnitude of 8 on the Richter scale, that is, an earthquake of a force that can be expected to occur once in 500 years (*Modern Power Systems* 1985).

There are some landslide problems upstream in the reservoir area, but these are not expected to constitute a danger to the dam. Seepage in a few karstic formations at the dam site, which was discovered during preliminary drilling, has been corrected by extending grout curtains toward the right and left abutments.

The Urfa Tunnel and Hilvan Canal
The Ataturk Dam is designed for electrical energy production and for supplying irrigation water to the Urfa Tunnel and the Hilvan Canal. Irrigation from the reservoir involves two separate systems

Table 2.9. Urfa Irrigation Tunnel[a]

Type	Circular, reinforced concrete lining
Length of Tunnel	26.4 km each, two parallel tunnels
Grade	$T_1 = 0.00062802$; $T_2 = 0.00062948$
Excavation Diameter	approximately 9.50 m
Completed Diameter	7.62 m
Concrete Lining	0.95–0.40 m
Amount of Rock Bolt	4,600,000 kg
Amount of Shotcrete	300,000 tons
Amount of Excavation	3,000,000 m^3
Amount of Concrete	1,150,000 m^3
Discharge of Tunnel	328 m^3/s
Area to Be Irrigated	300,000 ha
Geological Formation	Calcerous marl
Hydraulic Load	$T_1 = 40.25$ m; $T_2 = 39.74$ m

Source: DSI (1984b).

[a] The Urfa tunnel system consists of two concrete-lined tunnels, about 26 km long, through which water will be discharged from the Ataturk Dam. The water will be conveyed to a canal system at the beginning of the Urfa-Harran plains and will maintain the irrigation of approximately 300,000 ha.

because some plains which will receive water are either higher or lower than the reservoir's surface. The Urfa Tunnel system will take water from a branch of the reservoir near Bozova to irrigate about 300,000 ha of land by gravity flow. The Hilvan Canal system will use water pumped 107 m up from the reservoir; the intake is located upstream from the Urfa system and will provide water for about 400,000 ha. A total of 730,000 ha will be irrigated.

Construction on the Urfa Tunnel began in 1977 and was scheduled for completion in 1986 (DSI n.d.b), a date now postponed by unspecified delays. Water will pass through the two concrete-lined 7.5 m (inside diameter) tunnels for 26.4 km at a rate of 328 m^3/s (Table 2.9). Thereafter, canals, the first with a capacity of 53 m^3/s and the second with a 255 m^3/s capacity, will convey a yearly maximum of 9,700 Mm^3 of water to the Urfa-Harran and Mardin-Ceylanpinar plains.

The Hilvan Pumping Station will lift water from the Ataturk

reservoir to the Hilvan Canal. The canal will extend 150 km eastward and provide water for 64,500 ha in the southern part of the Siverek-Hilvan plains. A dam is to be built at Golebakan with a reservoir to supply water to secondary canals feeding another 79,500 ha in the same plains. The main canal will extend south and pass through the 5.7 km Siverek Tunnel. It will also pass through the 7.9 km Mardin Tunnel to convey water under the Tek Tek plateau to the Mardin plains. From that point, the canal will extend 81 km farther and irrigate another 140,000 ha. Pumping stations are planned to lift water to the Lake Yenice canal, thus supplying water to 113,000 ha located in the region's highest plains. When completed, the Hilvan system will be the largest single point source for any irrigation system in Turkey (DSI n.d.b).

The Tigris Portion of GAP

As a result of the emphasis placed on Euphrates development by Turkish planners, little is said concerning the eastern Tigris portion of the total project. Most of the discussion in this study will concentrate on the western river.

The Tigris portion of GAP will provide irrigation for about 600,000 ha of land and the generation of 8 billion kWh annually from an installed capacity of 2,000 MW of electricity (Ozis 1983). The plan calls for 4 large dams on the main Tigris, 3 on its tributaries, and 12 small pond (*golet*) dams on tributaries in irrigation areas. There are also 3 large, 4 medium, and 12 small power plants scheduled as well as numerous pumping stations.

In August 1985, DSI put out tenders for the TL 17.5 billion Kralkizi Dam (alternate spelling, Kiralkizi), hydroelectric station, and two associated tunnels (*Weekly Economic Report* n.d.). Data on this project and four other dams scheduled for the Tigris basin are given in Table 2.10.

Achieving the Goals of GAP

Achievement of the goals of the GAP cannot be realized without sometimes producing severe impact on the environment and the

Table 2.10. Dams and Hydroelectric Power Plants Planned and Under Construction in the Tigris River Basin in Turkey

Dam[a]	River	Province	Type	Embankment (Volume in Mm^3)	Reservoir (Volume in Mm^3)	(Area in sq km)	Irrigated (Area in ha)	Installed Power (Capacity in GW)
Kralkizi	Tigris	Diyarbakir	Earth/stone	14.953	1,919.0	57.50	80,000	56
Batman	Batman	Diyarbakir	Earth/stone	4.950	1,175.0	49.25	37,744	185
Dicle	Tigris	Diyarbakir	Stone	2.000	595.0	24.00	218,920	110
Ilisu	Tigris	Mardin	Earth/stone	33.500	10,410.0	299.50	—	1,200
Cizre	Tigris	Mardin	Stone	3.300	360.0	21,0	120,000	240

Source: DSI (1984a).

[a] The first three dams are under construction; the design for the Ilisu Dam is completed; the Cizre Dam is in the final design stage.

local populations. Such problems will be increased in the case of GAP because of its upstream position on two major international rivers. How much water will Turkey's irrigation schemes remove from the two rivers? How large will the return flow from the fields be, and what degree of water pollution will result? How will Turkey cope with the restructuring of the economy and society in its southeast? All such questions must be asked not only of Turkey and its development plans but also of Syria and Iraq farther downstream. Both of these countries entertain equally ambitious development programs which involve the use of the waters of the Euphrates and Tigris rivers. Certainly, any changes made anywhere along the length of either stream will have reverberations throughout the entire combined river system.

Not only *how* and *what* will happen are important questions that must be answered, but the timing of events is also critical. When will everything come on line? Will there be time to take remedial action, either technical or diplomatic, prior to crises that may arise? These and many other questions must be answered if the GAP is to realize its great potential in the most reasonable and effective way. It is the purpose of this work to analyze the conditions underlying such questions. Although this text does not purport to be a political analysis, the material found herein will be of great use to those who approach that difficult task.

3

Industry and Agriculture in the Southeastern Anatolia Project Region

Southeastern Anatolia presents something of an anomaly. It contains proportionately a third again as much good land as the national average in Turkey and is watered by two world-class river systems: the Tigris and the Euphrates. Yet it has for countless centuries been a remote backwater. It is sparsely populated, lacks the infrastructure one finds in other parts of Turkey, has less industry, and is less mechanized in agriculture.

The reasons for this anomaly are both physical and historical. The region is a backwater because it is, in fact, remote. The headwaters of the two river systems constitute a mountain fastness characterized in large part by rugged terrain protecting interior enclaves of rich valleys and plains. Historical activity has focused on the Bosporus and points west since Greco-Roman times. The Ottoman Empire was concerned first to conquer its Byzantine rival on the Golden Horn and then to establish the Sublime Porte to rule in its stead. Most economic, political and intellectual activity in the Ottoman Empire centered on Istanbul, and population followed suit. Thus, when the Turkish Republic succeeded the Ottoman Empire, it had first to develop those regions where the most people lived.

In consequence, modern Turkey finds in its southeastern region a truly neglected resource. It is an area with better than average land, abundant water, and rich mineral deposits. Development of this region has the potential to make Turkey a major exporter of agricultural products. Moreover, once the infrastructure is in place to improve the quality of life, the region could well become attractive and, to some extent, curb the rural-to-urban population migration that plagues developing lands. After all, southeastern Anatolia may have been a backwater for the last two thousand years, but for millennia before that it was a center of assorted Hittite, Hurrian, Mitannian, and Urartian civilizations.

Development Overview

The goals of the Southeast Anatolia Development Project (GAP) parallel those of Turkey's national Five-Year Plans. Since 1963, the government has encouraged development of the GAP region and other parts of Eastern Turkey. Goals to stop internal migration, to increase population, and to improve the standard of living in the region are documented in each of the subsequent Five-Year Development Plans. Government actions emphasize industrial investment using local agricultural products as raw materials. There is precedent for this in that investments in some manufacturing plants in the region create high amounts of value added although they are few in number and employ few workers. In 1982, two of the three large-scale establishments in Siirt province were state-owned factories, employing 3,525 people and creating value added of TL 20.2 billion. As a result, Siirt ranks twelfth in Turkey (Table 3.1) in value added by manufacturing.

In the most recent (the fifth) Five-Year Development Plan, the Southeastern and Eastern regions of Turkey are considered privileged and incentives are offered which encourage large-scale manufacturing plants. At the same time, small-scale labor-intensive establishments, particularly those highly sensitive to transport costs (such as dairy products), are also encouraged. The market region is considered to be the entire Middle East.

Table 3.1. Provinces Ranked by Industrial Establishments, Employees, and Value-Added, 1982

Rank	Number of Large Establishments		Average Number of Workers		Value-Added (Billion TL)	
1.	Istanbul	4596	Istanbul	275,038	Istanbul	540
2.	Izmir	840	Izmir	72,168	Izmir	203
3.	Ankara	636	Kocaeli	51,245	Icel	171
4.	Bursa	437	Ankara	46,203	Kocaeli	168
5.	Kocaeli	270	Adana	40,143	Ankara	71
6.	Adana	233	Bursa	40,049	Bursa	66
7.	Manisa	180	Zonguldak	24,123	Adana	60
8.	Konya	156	Hatay	19,461	Zonguldak	52
9.	*Gaziantep*	149	Konya	18,797	Tekirdag	21
10.	Denizli	128	Samsun	16,195	Samsun	21
11.	Eskisehir	122	Kayseri	14,561	Eskisehir	21
12.	Samsun	111	Eskisehir	14,311	*Siirt*	20
13.	Kayseri	110	Rize	14,285	Kayseri	19
14.	Aydin	107	Icel	14,135	Konya	18
15.	Zonguldak	105	Manisa	10,858	Balikesir	18
16.	Balikesir	105	Tekirdag	10,538	Samsun	13
17.	Icel	76	Sakarva	10,013	Manisa	11
18.	Sakarya	72	*Gaziantep*	9,133	*Gaziantep*	10
19.	Tekirdag	69	Balikesir	8,517	Antalya	9
20.	Antalya	60	Denizli	8,204	Edirne	9

Source: TSIS (1984).

Industry

Despite the encouragements and incentives offered in development plans, census statistics show that in absolute terms, the combined six provinces of the GAP region are still of minimal significance in terms of Turkey's industrial production. Only Diyarbakir and Gaziantep are important regional centers (Table 3.2), and even they provide only a small contribution to Turkey's industrial product.

Turkey conducted detailed censuses of its industries in 1950, 1963, 1970, and 1980. In addition to these four main industrial

Table 3.2. Functional Regions of Turkey

Regional Center	Provinces within Functional Region
Adana	Adana, Hatay, Icel
Ankara	Ankara, Cankiri, Corum, Kirsehir
Bursa	Bursa
Diyarbakir[a]	Diyarbakir, Bitlis, Hakkari, Mardin, Siirt, Van
Elazig	Elazig, Bingol Tunceli
Erzurum	Erzurum, Erzincan, Agri, Kars, Mus
Eskisehir	Eskisehir, Bilecik, Kutahya
Gaziantep[a]	Gaziantep, Adiyaman, Urfa, Kahramanmaras
Istanbul	Istanbul, Bolu, Canakkale, Edirne, Kirklareli, Kocaeli, Sakarya, Tekirdag, Zonguldak, Kastamonu
Izmir	Izmir, Afyon, Antalya, Aydin, Burdur, Denizli, Isparta, Manisa, Mugla, Usak, Balikesir
Kayseri	Kayseri, Nevsehir, Yozgat
Konya	Konya, Nigde
Malatya	Malatya
Samsun	Samsun, Amasya, Giresun, Ordu, Sinop, Tokat
Sivas	Sivas
Trabzon	Trabzon, Artvin, Rize, Gumushane

Source: *Turkiye'de Yerlesme Merkezlerinin Kademelenmesi Cilt I* (1982, Ankara).

[a] Within the GAP region.

census years, there have been intermediate years when production establishments with more than ten employees have been surveyed. All of these census and survey results are published by the State Institute of Statistics.

The six GAP provinces combined (Adiyaman, Diyarbakir, Gaziantep, Mardin, Siirt, Sanliurfa) account for only 1.85 percent of Turkey's 9,693 industrial establishments with more than ten employees. The six provinces have only 1.92 percent of Turkey's industrial employees and produce only 1.98 percent of the value added by large-scale manufacturing (firms with more than ten workers). At the other extreme, the most developed region of

Turkey is Istanbul. About 58 percent of all large-scale industries, 47.4 percent of industrial workers, and 44.6 percent of industrial value added are in the Istanbul region.

Within the GAP area, the two provinces of Gaziantep and Siirt rank high in importance for industry at the provincial scale. Gaziantep is ninth out of 67 provinces in number of establishments, eighteenth in terms of employees, and eighteenth in value added by manufacturing (Table 3.1). Gaziantep's proportion of economically active population employed in manufacturing industries, at 13.2 percent, is higher than Turkey's average of 10.7 percent. (This seeming contradiction demonstrates the concentration of industry in only a few provinces.) Siirt is not ranked in terms of establishments and workers but ranks twelfth out of 67 in terms of value added. It is in this favorable position because of petroleum and petroleum-related production. Oil fields near Batman account for significant population increases in the area— from 443 in 1945 to 86,172 in 1980.

Manufacturing industries in Turkey, with the exception of some heavy industries and State Economic Enterprises, have developed in response to market conditions. Consequently, industrial activities serving the undeveloped regions of Eastern and Southeastern Anatolia have agglomerated in the "functional service centers" of surrounding areas. Gaziantep has become such a functional center for the GAP region.

Gaziantep is one of only four "sixth grade" industrial centers in Turkey. (The others are Izmir, Ankara, and Adana; Istanbul, which exerts considerable influence in its region, is the only "seventh grade" center.) Gaziantep influences a vast area directly and indirectly and shares its influence with the functional center of Adana. The city of Diyarbakir has fewer large-scale manufacturing establishments than Gaziantep and is classified as a "fifth grade" functional center. Diyarbakir city serves the surrounding area as a center of health services, higher education, and administrative facilities.

As is generally true throughout Turkey, small-scale establishments (employing one to four persons) dominate the manufacturing sector in the six provinces of the GAP region. These provide

Table 3.3. Small-Scale Manufacturing Establishments—
Adiyaman Province, 1980

Province/ Districts	Number of Establishments	Number of Establishments by Number Employed				
		1	2	3–4	5–6	7–9
ADIYAMAN	577	272	167	114	19	5
Central District	278	112	87	61	14	4
Besni	92	41	31	18	1	1
Celikhan	15	9	2	4	—	—
Gerger	18	13	4	1	—	—
Golbasi	67	36	19	12	—	—
Kahta	107	61	24	18	4	—
ALL TURKEY	173,337	55,590	50,152	46,587	14,435	6,573

Source: TSIS (1982).

daily consumer goods and supply local demand (such as bakeries, tailors, brickmakers, and printing houses). Most are located in provincial and district centers, a few are scattered in rural settlements (Tables 3.3–3.8).

Only Gaziantep, out of the six GAP provinces, has significant numbers of large-scale manufacturing establishments, employing ten or more persons (Table 3.9). Large-scale industry in Gaziantep is mainly concerned with the processing of agricultural products. Such agro-industries include textile manufacture (depending on cotton), oil and soap production (depending on olives), and wine and raki (depending on grapes). Other units in Gaziantep city process foodstuffs, beverages, and consumer goods for local demand.

The basic industrial structure in the other provinces of the GAP region (Adiyaman, Diyarbakir, Mardin, and Sanliurfa) is similar but not identical to that of Gaziantep. Siirt is the only exception with its emphasis on petroleum production. All depend on the processing of local agricultural products, but the products differ. In Adiyaman, for example, one of the major industries is the processing of tobacco.

Table 3.4. Small-Scale Manufacturing Establishments—Diyarbakir Province, 1980

Province/ Districts	Number of Establishments	Number of Establishments by Number Employed				
		1	2	3–4	5–6	7–9
DIYARBAKIR	1,051	309	298	299	107	38
Central District	645	163	180	190	80	32
Bismil	72	22	22	25	2	1
Cermik	47	12	9	19	7	—
Cinar	22	8	9	5	—	—
Cungus	13	2	9	2	—	—
Dicle	8	2	3	3	—	—
Ergani	101	32	37	25	6	1
Hani	33	30	—	2	1	—
Hazro	20	6	10	4	—	—
Kulp	25	14	5	4	2	—
Lice	10	2	3	5	—	—
Silvan	55	16	11	15	9	4
ALL TURKEY	173,337	55,590	50,152	46,587	14,435	6,573

Source: TSIS (1982).

Table 3.5. Small-Scale Manufacturing Establishments—Gaziantep Province, 1980

Province/ Districts	Number of Establishments	Number of Establishments by Number Employed				
		1	2	3–4	5–6	7–9
GAZIANTEP	4,696	827	1,140	1,736	664	329
Central District	3,418	554	814	1,270	502	278
Araban	19	2	5	9	2	1
Islahiye	193	66	71	49	5	2
Kilis	614	135	128	212	106	33
Nizip	390	55	103	176	42	14
Oguzeli	47	14	14	17	1	1
Yavuzeli	15	1	5	3	6	—
ALL TURKEY	173,337	55,590	50,152	46,587	14,435	6,573

Source: TSIS (1982).

Table 3.6. Small-Scale Manufacturing Establishments—
Mardin Province, 1980

Province/ Districts	Number of Establishments	Number of Establishments by Number Employed				
		1	2	3–4	5–6	7–9
MARDIN	878	389	227	218	37	7
Central District	262	78	88	80	14	2
Cizre	129	65	31	30	3	—
Derik	10	2	5	2	1	—
Gercus	33	17	9	6	1	—
Idil	11	6	3	1	—	1
Kiziltepe	174	59	41	59	11	4
Mazidagi	12	5	4	3	—	—
Midyat	85	55	20	10	—	—
Nusaybin	101	80	10	6	5	—
Omerli	12	6	6	—	—	—
Savur	23	12	3	7	1	—
Silopi	26	4	7	14	1	—
ALL TURKEY	173,337	55,590	50,152	46,587	14,435	6,573

Source: TSIS (1982).

Potable Water for Domestic Use

An important adjunct of industrialization is the provision of a
suitable water supply for urban populations. In the GAP provinces,
water is obtained from different sources throughout the region.
Villages depend on local sources, however poor. In cities, serious
deficiencies require residents to transport water over long dis-
tances for both human and animal consumption. Major towns and
cities are facing increasing needs as their populations grow (Table
3.10).

Adiyaman's water requirement is supplied from the Kirkgoz
Springs about 10 km from the city. Due to a poor transfer system
only 0.22 Mm^3 is used annually from this source, although more
would be available if the system were improved.

Diyarbakir's potable water comes from the Gozelli Springs

Table 3.7. Small-Scale Manufacturing Establishments—
Sanliurfa Province, 1980

Province/ Districts	Number of Establishments	Number of Establishments by Number Employed				
		1	2	3–4	5–6	7–9
SANLIURFA	1,642	504	466	439	161	72
Central District	949	326	277	221	78	47
Akcakale	33	13	6	9	3	2
Birecik	138	19	44	55	17	3
Bozova	21	4	7	6	2	2
Halfeti	26	16	4	4	2	—
Hilvan	17	4	7	2	2	2
Siverek	164	58	39	43	21	3
Suruc	147	36	41	39	22	9
Viransehir	147	28	41	60	14	4
ALL TURKEY	173,337	55,590	50,152	46,587	14,435	6,573

Source: TSIS (1982).

Table 3.8. Small-Scale Manufacturing Establishments—Siirt
Province, 1980

Province/ Districts	Number of Establishments	Number of Establishments by Number Employed				
		1	2	3–4	5–6	7–9
SIIRT	564	201	183	140	28	12
Central District	226	96	83	40	6	1
Batman	226	55	61	78	21	11
Baykan	13	4	5	4	—	—
Besiri	13	6	6	1	—	—
Eruh	9	5	2	1	1	—
Kozluk	23	11	9	3	—	—
Kurtalan	33	16	12	5	—	—
Pervari	5	1	2	2	—	—
Sason	2	—	1	1	—	—
Sirnak	13	6	2	5	—	—
Sirvan	1	1	—	—	—	—
ALL TURKEY	173,337	55,590	50,152	46,587	14,435	6,573

Source: TSIS (1982).

Table 3.9. Large-Scale Industrial Establishments in GAP Provinces, 1964 and 1982

	Number of Establishments		Number of Workers		1982 Value Added (in 1,000 TL)
	1964	1982	1964	1982	
Adiyaman	0	3	0	1,227	1,546,671
Diyarbakir	6	14	766	1,537	1,546,734
Gaziantep	43	149	2,392	9,133	10,184,734
Mardin	0	2	0	32	14,378
Siirt	3	3	375	3,525	20,200,370
Sanliurfa	0	9	0	764	515,751
Regional Total	52	180	3,533	16,218	34,008,638
Regional Share	1.72%	1.85%	1.16%	1.92%	2.01%
TOTAL TURKEY	3,012	9,693	304,604	845,074	1,687,891,818

Source: TSIS (1964; 1982; 1984).

east of the city, which have a capacity of 340 l/s. Diyarbakir receives approximately 11 Mm^3 of water annually.

The city of Gaziantep obtains its water from the wells of the Pancarli Spring with a capacity of 250-300 l/s. The city receives 8.3 Mm^3 water annually.

In Mardin the city water is obtained from the Ayinseban Springs some 13 km from the city. The capacity of the springs is more than 50 l/s. However, due to a poor transfer system the city receives only a small part of that yield, approximately 0.3 Mm^3/yr.

Urfa's water is obtained from three different sources: the Kehriz Spring with a capacity of 15 l/s, the water from Fish Lake with a capacity of 145 l/s, and the Direkli Spring and wells, which supply the city with 6.24 Mm^3/yr.

Development plans for the GAP region include the improvement of urban water supplies (Table 3.11). However, the aggregate of such consumption is trivial compared to total usage, and the withdrawals from reservoirs for domestic use can be considered insignificant. Total domestic water use in the Euphrates basin after the year 2000 has been estimated by DSI at approximately 92.5 Mm^3/yr (DSI 1984b, 27). The same source anticipates 82.5 Mm^3 will be used for industrial purposes in the GAP region.

Table 3.10. Existing Water Consumption in Provincial Centers of GAP, 1985
(liters/person/day)

Province	Domestic Use	Industrial Use	Total
Adiyaman	18	—	18
Diyarbakir	180	30	210
Gaziantep	151	17	168
Mardin	20	—	20
Urfa	128	20	148
Siirt	NA	NA	NA

Source: Mehmet Tomanbay, *Administrative and economic parameters of water use in the Southeast Anatolia Project,* in ms. (Sept 1988) as selected from appendices found in EMSA (Etud, Musavirlik Sanayi Arastirmasi A.S.), *Guneydogu Anadolu Bolgesi Gelisme Plani,* Istabul, Kutlu Ozalit (1987) and DSI, *Icme Sulari ve Kanalizasyon Dair Baskanligi, Icme Sulari Kesin Reporlari.*

Agriculture in the GAP Region

All six of the provinces of the GAP region will feel the results of increased irrigation. Many areas will benefit in Urfa, Mardin, and Diyarbakir, while fewer areas will be affected in Adiyaman, Gaziantep, and Siirt. Since Turkish statistics are published by prov-

Table 3.11. Future Water Needs of Provincial Centers
(consumption in liters/person/day)

Province	Year	Domestic	Industrial	Total
Adiyaman	2000	150	20	170
Mardin	2020	—	—	250
Siirt	2022	—	—	200
Gaziantep	2020	231	25	256
Diyarbakir	2025	216	(383 l/s)	(5,380 l/s)
Sanliurfa	2020	290	67	357

Source: Mehmet Tomanbay, *Administrative and economic parameters of water use in the Southeast Anatolia Project,* in ms. (Sept 1988) as selected from appendices found in DSI, *Icme Sulari ve Kanalizasyon Dair Baskanligi, Icme Sulari Kesin Reporlari,* and Iller Bankasi, *Icme Sulari Dairesi Baskanligi* (ca 1985).

Table 3.12. Employment of Economically Active Population by Economic Sector in Provinces of the GAP Region, 1980
(percent of economically active population employed in each)

Sectors	Adiyaman	Diyarbakir	Gaziantep	Mardin	Sanliurfa	Siirt
Agriculture	82.7	71.8	53.0	77.6	76.7	70.7
Manufacturing	3.7	3.0	13.2	2.1	2.8	3.6
Social Services	6.1	13.4	13.6	9.9	2.7	11.3
Commerce	2.6	4.4	8.7	2.7	2.7	11.3
Construction/ Building	2.3	4.4	4.9	2.5	2.4	2.7

Source: TSIS (1983a).

inces, the analysis of agriculture in the GAP region will be presented at the provincial level. It is not possible with current data to disaggregate the subregions of the GAP program.

With the exception of Gaziantep, the economies of all GAP provinces rely largely on agriculture. The region is distant from the more developed western part of Turkey and has a much lower population density than Turkey as a whole. Population density per sq km ranges from Sanliurfa (31), Siirt (36), Mardin (44), Adiyaman (47) and Diyarbakir (51), which is close to but still less than the national norm, to Gaziantep (98), almost twice the national average (56).

Arable and Cultivated Land
The agricultural nature of the region is clearly displayed in Table 3.12. With the exception of Gaziantep, about three-fourths of the active population is employed in agriculture raising a variety of crops (Table 3.13).

Agriculturally, all the GAP provinces show common characteristics, but there are some differences. The percentage of arable land varies from 9.5 in Siirt to 49.5 in Sanliurfa (Table 3.14). This is significant because the new GAP irrigation is closely linked to the percentage of Class I, II, III arable land combined with the Class IV restricted arable lands (see Chapter 1, Land management,

Table 3.13. Types of Agricultural Crops in GAP Region

Type of Crop	Crops
Cereal	Wheat, barley, maize, millet, rice
Pulses	Chick peas, dry beans, lentils, mungi beans
Industrial crops	Tobacco, sugar beets, cotton (lint), flax (fiber), dry pepper
Oil seeds	Cotton seed, sesame, flax (seed), soy beans, sunflowers, groundnuts
Tuber crops	Dry onions, dry garlic, potatoes
Fruits	Pears, quinces, apples, plums, apricots, cherries, peaches, sour cherries, wild apricots, olives
Nuts	Pistachios, walnuts, almonds
Grape-like fruits	Mulberries, figs, pomegranates, grapes
Fodder crops	Maize, cow vetch, wild vetch, alfalfa, sainfoin

Source: TSIS, *Agriculture Structure and Productivity* (1982).

classification, and drainage). For Sanliurfa, the 12.7 percent of Class IV land increases its total arable land to 62.2 percent; thus the Harran Ova irrigation project in Sanliurfa province appears promising. Except for Siirt and Adayaman provinces, which will only in part be affected by GAP, all of the GAP region provinces have a higher rate of arable land than the 25.2 percent average for Turkey.

The percentage of arable land cultivated also differs by province. In 1982, 97.0 percent of the arable land in Adiyaman was cultivated. In the same year, the rate was 92.4 percent for Diyarbakir, 87.1 percent for Mardin, 81.6 percent for Gaziantep, and 75.4 percent for Sanliurfa. Siirt pushed beyond the normal limits of its arable land, actually cultivating 12.4 percent of its land classified as Class IV (restricted arable) (Table 3.15).

Agricultural Land Use and Crops
The GAP region is an area with a dry season lasting up to six months out of each year. Because of this aridity, traditional dry farming methods—primarily idle fallow—are practiced (Tables

Table 3.14. Land Classification in the GAP Provinces
(in hectares)

| Province | Arable Land by Class in Hectares | | | Total Arable Land | | Restricted | Non-Arable | Non Agric. | Total |
	Class I	Class II	Class III	Hectares	Percent	Class IV	V-VI-VII	Class VIII	
Adiyaman	21,928	58,258	64,432	144,618	19.1	56,733	520,332	36,616	758,299
Diyarbakir	150,765	225,104	180,031	555,900	36.3	140,468	781,150	52,728	1,530,246
Gaziantep	80,753	122,939	76,224	279,916	36.6	86,572	389,717	6,910	763,115
Mardin	181,598	137,788	120,367	439,753	35.7	95,518	640,490	55,169	1,230,930
Siirt	22,489	39,698	41,939	104,126	9.5	37,702	870,258	84,163	1,096,249
Sanliurfa	454,219	234,903	249,572	938,694	49.5	239,998	698,737	19,204	1,896,633
Reg. Total	911,752	818,690	732,565	2,463,007	33.9	656,991	3,900,684	254,790	7,275,472
Reg. Share	18.2%	12.1%	9.7%	12.7%		9.1%	8.4%	7.4%	9.5%
All Turkey	5,012,537	6,758,702	7,574,330	19,345,569	25.2	7,201,016	46,692,633	3,455,513	76,694,731

Source: Topraksu Genel Mudurlugu. n.d. *Turkiye Arazi Varligi: Kullanma, Siniflar, Sorunlar*, 28–29.

Table 3.15. Arable and Cultivated Land in GAP Region

Province	Arable Land	Cultivated Land	Cultivated Land as Percent of Arable
Adiyaman	144,618	140,305	97.0
Diyarbakir	555,900	513,623	92.4
Gaziantep	279,916	228,304	81.6
Mardin	439,753	382,994	87.1
Siirt	104,126	117,079	112.4
Sanliurfa	938,694	708,047	75.4
REGION	2,463,007	2,090,352	84.9

Source: *Tarimsal Yapi ve Uretim* (1982).

3.16–3.19). The largest areas are sown in wheat and barley, cereals resistant to drought and suitable for dry farming. There are differences in the productivity of the provinces. In Siirt, the area devoted to cereals (88.5 percent) is above the national average (80.1 percent), but in the other five provinces the proportion of land devoted to cereals is lower than the national average. Irrigation practices in some parts of Gaziantep and Mardin favor other crops over cereals (Tables 3.20–3.21). Irrigated cotton (lint and seed) and grape production in Mardin and Gaziantep compete for land use with cereals.

Nevertheless, wheat is the major cereal crop in southeastern Anatolia. For this area, except in Diyarbakir province, wheat comprises a higher percentage of cereal than for Turkey as a whole. Barley is the second most important crop in terms of area sown.

Production

Provinces in the GAP region vary in terms of crop yields and area cropped. Collectively, wheat represents 71.1 percent of the total cereal production and occupies 70.1 percent of the area planted in cereals. The six provinces contribute only 9.5 percent of Turkey's total cereal production. For the two major cereals, the GAP produces 10.2 percent of the nation's wheat and 11.1 percent of the barley.

However, for some crops the GAP region is nationally significant. The six provinces produce 74.6 percent of Turkey's lentils

Table 3.16. Major Production Groups in GAP Region
(area in hectares)

Province	Cereals		Industrial Crops		Pulses		Tuber Crops		Oil seeds	
	Area	Percent	Area	Percent	Area	Percent	Area	Percent	Area	Percent
Adiyaman	110,303	78.6	11,420	8.1	16,420	11.7	657	0.5	1,505	1.1
Diyarbakir	369,642	72.0	7,721	1.5	131,145	25.5	1,540	0.3	3,575	0.7
Gaziantep	145,133	63.6	15,818	6.9	61,743	27.0	3,205	1.4	2,405	1.1
Mardin	248,041	64.8	9,269	2.4	123,494	32.2	1,140	0.3	1,050	0.3
Siirt	103,701	88.5	5,906	5.0	6,857	5.9	537	0.5	78	0.1
Sanliurfa	512,834	72.4	11,198	1.6	178,604	25.2	682	0.1	4,729	0.7
REGION	1,489,654	71.3	61,332	2.9	518,263	24.8	7,761	0.4	13,342	0.6

Region: Total area cropped = 2,090,352 ha

Source: *Tarimsal Yapi ve Uretim* (1982).

Table 3.17. Cereal Production in GAP Region
(all cereals)

Provinces	Area in hectares	Production in tons
Adiyaman	110,303	179,858
Diyarbakir	369,642	579,210
Gaziantep	145,133	273,749
Mardin	248,041	595,982
Siirt	103,701	145,106
Sanliurfa	512,834	741,729
REGION	1,489,654	2,515,634
Turkey	13,421,685	26,418,000

Source: *Tarimsal Yapi ve Uretim* (1982).

(410,457 tons out of a total of 500,000 tons). The provinces of Gaziantep and Sanliurfa on the average produce 88.2 percent of Turkey's pistachios. (In fact, one name used for pistachio in Turkey is "Antep fistigi," literally "the nut of Gaziantep," Antep being a historical name for Gaziantep.) The quantity of pistachios produced and shares by province do vary from year to year; in 1982 Gaziantep contributed 17.3 percent of Turkey's total yield, Sanliurfa 54.6 percent. Grapes are also important. The GAP region produces over one-fifth of Turkey's total grape production. In 1982, out of a regional total of 821,424 tons, Gaziantep produced 405,000 tons, Diyarbakir 117,150 tons, and Adiyaman 102,150 tons.

The industrial crops of cotton lint, tobacco, and sugar beets are grown mostly in the irrigated areas. Sugar beets and tobacco are produced in Adiyaman; cotton in Mardin, Diyarbakir, Sanliurfa, and Gaziantep; and tobacco in Siirt.

Other crops that rank nationally in importance include sesame, almonds, and pomegranates. Sanliurfa produces 9.9 percent of Turkey's 27,000 tons of sesame. During the seventeenth century, Harran Ova shipped sesame as far as Marseilles and Geneva. Mardin produces 14 percent of Turkey's almonds; Siirt provides 15.2 percent of the nation's pomegranates.

For vegetables, the GAP region claims 10.4 percent (64,239 ha) of the total area devoted to those crops (Table 3.20). In 1982,

Table 3.18. Wheat Production in GAP Region

Province	Area in ha	% of Cereals	Production in tons	% of Cereals	Yield in Kg/ha
Adiyaman	79,881	72.4	125,115	69.5	1,566
Diyarbakir	235,500	63.7	343,375	59.3	1,458
Gaziantep	107,057	73.8	197,362	72.1	1,844
Mardin	191,213	77.1	472,396	79.3	2,471
Siirt	80,870	78.0	112,203	77.7	1,387
Saniliurfa	360,710	70.3	537,880	72.5	1,491
REGION	1,055,231		1,788,331		
TURKEY	9,000,000	67.1	17,500,000	66.2	1,944

Source: *Tarimsal Yapi ve Uretim* (1982).

the GAP region produced 8.5 percent (1,054,815 tons) of Turkey's total of 12,420,908 tons of vegetables. Although reliable statistics on vegetable production are fragmentary, there does seem to be an increasing trend of production in the GAP region. At present, with a few exceptions, vegetables are mainly raised for local consumption. The most important exceptions are the "fruit bearing vegetables"—melons and watermelons. The six GAP provinces produced 15.5 percent of Turkey's watermelons and 14 percent of

Table 3.19. Barley Production in GAP Region

Province	Area in ha	% of Cereals	Production in tons	% of Cereals	Yield in kg/ha
Adiyaman	29,213	26.5	56,651	29.3	1,939
Diyarbakir	123,700	33.5	219,780	37.9	1,777
Gaziantep	37,316	25.7	74,252	27.1	1,990
Mardin	56,478	22.8	122,935	20.6	2,177
Siirt	22,540	21.7	31,794	21.9	1,411
Saniliurfa	151,640	29.6	203,290	27.4	1,341
REGION	420,887		708,702		
TURKEY	3,137,000	23.4	6,400,000	24.2	2,040

Source: *Tarimsal Yapi ve Uretim* (1982).

Table 3.20. Vegetable Production in GAP Region

Province	Area in ha	Product in tons	Fruit Bearing Vegetables	Percent
Adiyaman	6,202	78,773	76,912	97.6
Diyarbakir	19,747	384,534	381,014	99.1
Gaziantep	9,816	241,353	233,426	96.7
Mardin	13,288	197,658	194,537	98.4
Siirt	3,766	78,789	76,425	97.0
Sanliurfa	11,420	63,708	69,651	94.5
REGION	64,239	1,054,815	1,031,965	97.8

Source: *Tarimsal Yapi ve Uretim* (1982).

its melons in 1982; Diyarbakir province was fourth and Mardin seventh in watermelon production.

It is difficult to make precise predictions regarding the impact of GAP on the production of crops both within the region and throughout Turkey. Detailed soil analyses, prevailing market conditions, and the skills and predilections of farmers/investors will all help to determine what eventually is produced on the irrigated lands of the project. Nevertheless, a number of estimates of future

Table 3.21. Irrigated Land in GAP Region
(public and private)

Province	Area in ha
Adiyaman	12,336
Diyarbakir	26,114
Gaziantep	22,294
Mardin	22,256
Siirt	5,060
Sanliurfa	33,694
TOTAL irrigated in GAP region	121,754
TOTAL irrigated in Turkey	2,990,080
Additional land to be irrigated by the Southeast Anatolia Project	1,800,000

Source: GAP (1980).

Table 3.22. Anticipated Crop Production Resulting from GAP
(in 000 tons)

Crop	Turkish Production[a] Before GAP	Anticipated Addition by GAP Development
Cotton (ginned)	475	556
Cotton (unginned)	NG	1,391
Oil seed (including cotton seed)	1,274	1,107
Oil seed (excluding cotton seed)	NG	273
Rice (in husk)	NG	515
Rice (milled)	190	400
Sugar beet	8,837	4,269
Tubers (including sugar beets)	NG	4,289
Grains	22,750	2,467
Clover/alfalfa	631	1,254
Fodder	NG	2,692
Vegetables/melons	9,265	3,033
Onions	NG	46
Legumes	NG	83.5
Tobacco	NG	18
Grapes	3,496	406
Fruit	1,253	390
Olives	NG	8
Pistachios	NG	9.9
Various crops	NG	270
Poplar (wood in 000 m^3)	NG	233

Source: GAP (1980).

[a] Year unspecified but prior to 1980.
Note: NG=Not given in quoted sources. No attempt is made to supply figures from other sources.

production resulting from GAP irrigation are available. Two such sets of data— apparently stemming from one original but unidentified source—are shown in Table 3.22. Cotton, oil seed (including cotton seed oil), sugar beets, grains, clover and alfalfa, fodder crops, and melons (which will triple in quantity) are all given importance by these data. Not only marketing demand but also marketing

Table 3.23. Number of Tractors in GAP Provinces

Province	1976	1978	1982	Percent Increase
Adiyaman	1,514	1,761	2,432	61%
Diyarbakir	2,784	3,517	4,585	65%
Gaziantep	2,644	3,611	6,818	157%
Mardin	1,288	1,869	2,150	67%
Siirt	560	974	1,227	119%
Sanliurfa	2,519	2,969	4,085	62%
REGIONAL TOTAL	11,309	14,701	21,297	88%
ALL TURKEY	281,802	370,259	491,001	74%

Source: GAP (1980, p. VII–10) for 1976 and 1978 statistics. 1982 statistics are from *Tarimsal Yapi ve Uretim* (1982).

Note that land, equipment, and crops occur in these six provinces outside as well as inside the GAP development area.

ability will be of great importance if these quantities are to be sold profitably, if and when they are produced. (Further comments on this situation are presented in Chapter 11.)

Mechanization

Southeastern Anatolia, including the area of GAP development, has lagged behind the rest of Turkey in terms of agricultural mechanization. In absolute numbers, by 1982 there were 21,297 four-wheeled tractors in the six GAP provinces, compared with 497,000 tractors in Turkey as a whole (Table 3.23). A recent study shows the disproportion between the southeast and all of Turkey with regard to horsepower and area cultivated (Table 3.24). In the southeast there was less than 0.5 horsepower per hectare of land available, compared with 1.05 horsepower per hectare for the total country. By the same token, each tractor in the southeast was on the average used to cultivate two and two-thirds more land than elsewhere in Turkey. The same situation applied to tractor-drawn equipment. Moldboard plows, disc harrows, harvesters, and farm wagons were all less abundant in the southeast than in the entire country (Table 3.25).

The growing importance of mechanization to southeastern

Table 3.24. Levels of Farm Mechanization in Turkey, 1983[a]

Region	No. Tractors	Area Cultivated	Horsepower	HP/ha	Tractors/ha	ha/Tractor
Southeast Anatolia	22,249	2,683,889	1,124,291	0.42	0.00829	120.6
All Turkey	512,275	22,972,000	24,241,663	1.05	0.02230	44.8

Source: Gungor Yavuzcan et al. (1986, 453–467).

[a] Although the source cites these as appearing in a 1982 report, the number of tractors corresponds to 1983.

Table 3.25. Pieces of Equipment per Tractor in Turkey

Type of Equipment	1980		1984	
	Turkey	SE Anatolia	Turkey	SE Anatolia
Moldboard plow	0.89	0.64	0.85	0.73
Cultivator	0.34	0.56	0.38	0.64
Disc harrow	0.20	0.07	0.18	0.13
Seeder	0.20	0.33	0.20	0.28
Universal harvester	0.05	0.02	0.05	0.01
Hay mower	0.02	0.01	0.02	0.04
Wagon	0.97	0.88	0.89	0.78

Source: Gungor Yavuzcan et al. (1986, 453–467).

farming is indicated by the rapid increase in tractors and equipment in that region. The number of tractors—while still below the national average—increased 88 percent from 1976 to 1982 in the six GAP provinces, while only growing by 74 percent elsewhere. Nevertheless, the intensification of agriculture anticipated in the GAP area will necessitate greater advances. Energy in the form of fuel and electricity must also be increased. In the latter case, hydropower generation at the many GAP plants should meet future needs. Petrofuels will have to be imported to the region at an increasing rate. Meanwhile, village electrification in 1985 was 84.6 percent complete throughout Turkey but included only 66.8 percent of the southeast region's villages. Similarly, in 1985 60 kWh/ha/yr of electricity was expended for Turkey as a whole but only 23.4 kWh/ha/yr was expended in the southeast (Girgin et al. 1986).

It remains to be seen what efforts the government will make to meet growing demands for equipment and energy. In May 1987, a contract for $3.2 million was given to erect an office building in Urfa to house the 92 social projects thus far drawn up to meet GAP objectives (*MEED* 1987c). This indicates Turkish attention to the development of economic and social infrastructures in the GAP region. Yet unanswered is the question of how such change will take place; for instance, will tractor cooperatives be formed or will equipment be provided on the basis of private ownership?

Table 3.26. Number of Agricultural Holdings in GAP Region

Provinces	Farms with Crops Only		Farms with Crops & Animals		Total Number
	Number	Percent	Number	Percent	
Adiyaman	2,213	7.3	28,055	92.7	30,268
Diyarbakir	1,699	4.0	40,267	96.0	41,966
Gaziantep	3,933	12.3	27,902	87.7	31,835
Mardin	4,239	10.1	37,622	89.9	41,861
Siirt	2,056	8.7	21,639	91.3	23,695
Sanliurfa	3,922	12.5	27,333	87.5	31,255
TOTAL	18,062	9.0	182,818	91.0	200,880

Source: GAP (1980, VII–5).

Animal Husbandry

The two types of agricultural holdings in the GAP region are: (1) those used only for crop production, and (2) those used for both crop production and animal husbandry. Some 91 percent of agricultural holdings in the six GAP provinces are of the latter type (Table 3.26). Because agricultural productivity is low due to lack of extensive irrigation and mechanization, husbandry is an essential supplement to agriculture. However, poor pasturage and water deficiencies have a negative influence on husbandry.

Considering only adult animals of the species that are the major source of meat throughout Turkey (sheep, goats, angora goats, cattle), 12.4 percent of Turkey's 82,333,000 head of such livestock were in the six GAP provinces in 1982, while 11.1 percent of the nation's livestock were slaughtered and consumed there. Sheep and goats, ordinary and angora, are the main animals in the region (84.3 percent). Sanliurfa led the provinces with 2,562,360 animals, followed by Mardin with 2,277,563, and Siirt and Diyarbakir with 1,859,951 and 1,847,949, respectively. Diyarbakir province led the other five in cattle production; the six provinces together raised 7.1 percent of Turkey's cattle in 1982 (Table 3.27).

Although the intensification of farm methods associated with irrigation usually precludes animal husbandry, the inclusion of

Table 3.27. Number of Animals in GAP Provinces, 1982

Province	Total	Sheep	Goats	Angora Goats	Cattle	Animals Slaughtered
Adiyaman	806,219	383,460	313,510	0	109,249	67,980
Diyarbakir	1,847,949	837,130	584,700	0	426,119	295,690
Gaziantep	842,391	438,620	340,270	0	63,501	317,990
Mardin	2,277,563	1,221,470	486,110	404,460	165,523	146,050
Siirt	1,859,951	807,250	742,610	174,530	135,561	135,380
Sanliurfa	2,562,360	2,122,380	315,720	0	124,260	151,044
REGIONAL TOTAL	10,196,433	5,810,310	2,782,920	578,990	1,024,213	1,114,134
TURKEY	82,333,000	49,636,000	14,655,000	3,558,000	14,484,000	10,055,000

Source: TSIS (1988, 205, Tables 167–168).

Note: Data for slaughtered animals derived from municipal slaughterhouses. Data do not include animals killed illegally or for sacrifice. Animals slaughtered may include other species, e.g., water buffalo.

Table 3.28. Categories of Landholding in GAP Region

Farm Size in ha	Number of Families	Aggregate Area in ha	Percent of Owners in Category	Percent of Farmland in Category
1.0–5.0	141,903	199,075	61.4	10.5
5.1–20.0	74,843	756,291	32/4	40.0
20.1–50.0	12,211	395,559	5.3	20.9
50.1–100.0	603	66,880	0.3	3.5
100.1+	1,389	473,787	0.6	25.1
TOTAL	230,949	1,891,592	100.0	100.0

Source: GAP (1980, VIII–5).

large amounts of fodder and clover in projected crop goals (Table 3.22) would indicate that the number of animals raised in the GAP area should, if anything, increase in the future.

Land Ownership

Well over half of the agricultural land holdings in the GAP provinces range between 1 and 5 ha in size, but these small holdings comprise only 10.5 percent of the cultivated lands in the region. On the other hand, although less than 1 percent of the holdings are considered large (more than 100 ha), they take up a quarter of the cultivated land. These facts point up a considerable inequality in the region's land ownership patterns (Table 3.28).

At this writing, much more needs to be known about the distribution of land by owner and farm operation in the GAP area. Several small studies on the subject have been undertaken by the Turkish Foundation for Scientific and Technical Research (Turkiye Bilimsel ve Teknik Arastirma Grubu/TUBITAK). Table 3.29 shows the size and ownership of a small area (76.28 ha) in Urfa Vilayet sampled for discussion at a 1986 symposium on the agricultural development of the GAP area (Girgin et al. 1986).

It is difficult to compare these figures with those in Table 3.28, but a pattern does emerge: a relatively small portion of the area is in very small holdings and a disproportionately large portion is contained in larger properties. Nearly two-thirds of the sampled holdings were in operations more than 25.1 ha in size.

Table 3.29. Size and Ownership in a Sample of Agricultural Enterprises (Working Farms) in Urfa Province
(in decares: 10 da=1 ha)

Size of Operation in da	Area Farmed by Form of Land Tenure						Total Area in da	Percent of Total
	Private		Rented		Share-cropped			
	da	%	da	%	da	%		
1–50	16.7	46.4	16.7	46.4	2.6	7.2	36.0	4.7
51–100	48.8	61.9	18.9	24.0	11.1	14.1	78.8	10.3
101–250	88.2	53.9	67.4	41.2	7.9	4.9	163.5	21.4
251 or more	310.6	64.1	155.7	32.1	18.2	3.8	484.5	63.6
TOTAL	464.3	60.9	258.7	33.9	39.8	5.2	762.8	100.0

Source: Aksoy et al. (51, Table 4).

Table 3.30. Sample of the Size of Farming Operations in
Merkez and Akcakale Ilceler, Urfa Province

| Size of Operation | Farm Families | | Area Farmed | |
in da	Number	Percent	Decares	Percent
24–50	19	21.1	684	4.5
51–100	19	21.1	1,498	9.9
101–250	38	42.2	6,215	40.9
251–1570	14	15.6	6,783	44.7
TOTAL	90	100.0	15,180	100.0

Source: Aksoy et al. (53, Table 5).

Almost the same percentage of the total land was held in large,
privately owned parcels. Sixty-one percent of holdings of all sizes
are privately owned, 34 percent are rented, and 5 percent are
farmed by sharecropping.

According to the authors of this study, the percentage of
renting is higher in the GAP area than the Turkish average but
sharecropping is considerably lower. The latter is true because,
contrary to the usual pattern of sharecropping in which poor farmers supply labor and owners provide soil and seed, in the GAP
region owners provide soil and seed while contract laborers provide tractors and equipment as well as work, for half the subsequent harvest (Girgin et al. 1986). Rented land is more frequent
in the GAP region because of the expropriation of some lands by
the Secretariat for Soil and Agricultural Reform (Toprak ve Tarim
Reformu Mustesarligi/TTRM). In 1985, TTRM is reported to have
rented 352,363.9 ha (Girgin et al. 1986), but it is unclear to whom
these lands were let.

Table 3.30 is based on a sample of 90 owner-operated farms.
It shows 21 percent of the farm families operating plots of less
than 5 ha, a total of 68.4 ha or 4.5 percent of the land. Nearly half
the land (44.7 percent) was farmed by only 14 families (15.6 percent
of the total number of families). It should be born in mind that
landless families represent a significant portion of the total rural
population in the GAP area. Table 3.31 shows that in Urfa Vilayet,
42.3 percent of the rural population was landless; in the central
Ilce (*kaza* or county) of the province, the proportion of landless

Table 3.31. Distribution of Landless Familes in Urfa Il (689 Villages) and Merkez and Akcakale Ilceler, 1981
(village inventory study)

Il/Ilce	Total Farm Families		Landowning Families		Landless Families	
	Number	Percent	Number	Percent	Number	Percent
Urfa	73,579	100.0	42,433	57.7	31,146	42.3
Merkez	14,727	100.0	7,446	50.6	7,281	49.4
Akcakale	9,484	100.0	4,969	52.4	4,515	47.6

Source: Aksoy et al. (53, Table 6).

rural population was 49.4 percent. According to a 1975 study (Aresvik 1975, 37–38), 16.5 percent of the farmers were landless in all of Turkey.

What emerges is a picture of large landholdings in the hands of relatively few people and a great number of landless farm folk working for others. Whether or not land reform will ameliorate this situation—or simply end up with land being rented to wealthy tractor owners—is unclear at present. In any event, much more effort must be made by both researchers and the government to address and correct this situation, which in view of the low standard of living in southeastern Turkey will remain politically sensitive as GAP development matures.

Prospects

It is too early to forecast the probable effects of GAP development on the region. Indeed, the necessary background studies are not yet complete. The original "Lower Euphrates Project" aimed to build dams in an area considerably smaller than today's GAP; hence land and soil classifications have been done for only a part of the present project region. A four-year study was begun in 1985 to classify the soils in the whole GAP area, on the basis of which planners can determine which crops to grow. It is likely that traditional crops will continue to be grown but that more efficient methods and dependable water supplies will produce significantly higher yields. Continuation of traditional crops will minimize the changes required in social and economic structures.

A considerable amount of land will be inundated by the new reservoirs. These alluvial soils, among the region's most fertile, will be replaced by expansion of cultivation into "restricted or marginal arable lands" (Class IV). High yields on these lands can only be achieved through irrigation, fertilization, and even more advanced techniques. According to plans, introduction of irrigation will increase the importance of industrial crops, which will depend closely on the demands of planned new manufacturing plants. Market conditions will play an important role in deciding what crops will be sown. At present these conditions are far from stable; base prices and production quotas are very unpredictable in Turkey.

Mechanization of agriculture will result in increased yields of crops but will significantly increase the demand for equipment. Since a tractor plows about 45 ha of land, it will require more than 33,000 tractors to cultivate the more than 1.5 million ha of land to be irrigated (a 55 percent rise from 1982's total of 21,297 tractors in the six provinces). Increases in crop yields will be substantial. For example, alfalfa production is expected to reach 1.25 million tons after GAP is completed, more than twice the present production level for all of Turkey. Animal husbandry is also expected to improve.

Expanded irrigation and new power supplies will bring a vast spectrum of social and economic challenges to the GAP region. Farmers will have to be trained in irrigation methods, the use of fertilizers, and advanced agricultural technologies. Institutions such as DSI and the Ministry of Agriculture will have major roles in the educational programs that will begin after the first few GAP sub-projects come on line. The mere size of the GAP project gives the attendant training and infrastructural changes a major social significance.

A further note should be made regarding the concern for the environment which must attend any project of such magnitude. Questions regarding environmental impact are asked by the newly formed Turkish Environmental Issues Foundation (Turkiye Cevre

Sorunlari Vakfi, or TCSV). Created by law Number 2872 in August 1983, this group is beginning to address some of the environmental issues raised by GAP developments. It is premature to speculate on the effectiveness of this group, but its actions will warrant future assessment.

4

The Impact of Development
upon the Waters of the Euphrates River
and Its Tributaries

The preceding overview of the Southeast Anatolia Development Project (GAP) leads inevitably to the focus of this study. That is, what will be the impact on the three riparian users from developments—both in place and planned—on the Euphrates River in Turkey and in Syria? These questions for the Tigris River are at this juncture less critical for two reasons: Turkish development of the Tigris and its basin has scarcely begun, and the regime of the Tigris downstream in Iraq presents special problems unlike those relating to the Euphrates. This latter condition results from the Tigris' receiving large increments of water from left bank tributaries throughout its course in Iraq, while the Euphrates is an exotic stream in Iraq and even in Syria is far more dependent upon Turkish sources than usually thought.

To address the above question, we must review the Euphrates River from its source to its mouth in terms of dams, reservoirs, and diversions for irrigation as well as of evaporation, evapotranspiration, water losses, and return flows. Not all readers may have a generalized view of how human use of a river system impacts upon the riverain environment. Thus, a brief review of this topic, which can be skimmed or skipped without loss by the cognoscenti, is found in this chapter.

Such a review emphasizes two sets of problems. The first is

to define the above terms as they are used or can be expected to describe activities and phenomena referred to in the various articles and technical reports upon which this commentary is based, for it is necessary to use both Turkish and Syrian materials, references, and research. In this way, the results garnered from each country cross-check the results from the other and may serve to fill in gaps in the data by measurement at another venue or by extrapolation. The second problem refers specifically to the spotty and less than complete information on certain aspects of Syrian Euphrates development activities which must be worked out in detail in the pages ahead before the total review referred to above can be attempted. To clarify the interrelationship of these elements, we must consider a general model of river use in the Middle East (Figure 4.1).

Characteristics of Middle East River System Use

Streams in the Middle East are largely "exotic" by nature; that is, they rise in well-watered areas but before reaching the sea or some inland sink they flow into an arid zone where no more water is added and they actually diminish in volume through evaporation and seepage, not to mention human use. The basic characteristic of such streams is that they have seasonal periods of high water followed by periods of extremely reduced flow. For example, whereas the St. Lawrence River has only twice as much water at high flow as at low flow, the Nile has more than eight times as much water in September as in May, the Euphrates 28 times its minimum amount, and the Tigris nearly 80 times as much. Such flows are the result of winter rains in higher areas, the melting of the mountain snow pack, or, in the case of the Nile, the onslaught of the monsoon onto the Ethiopian highlands.

There are at least six uses for such rivers. In approximate diminishing order of importance these are: irrigation, domestic use, hydropower, industrial use, navigation, and fisheries. The latter two uses are eclipsed by the first four, of which hydropower is the least demanding; use of river waters to generate power usually does not deplete or change them. There are two exceptions

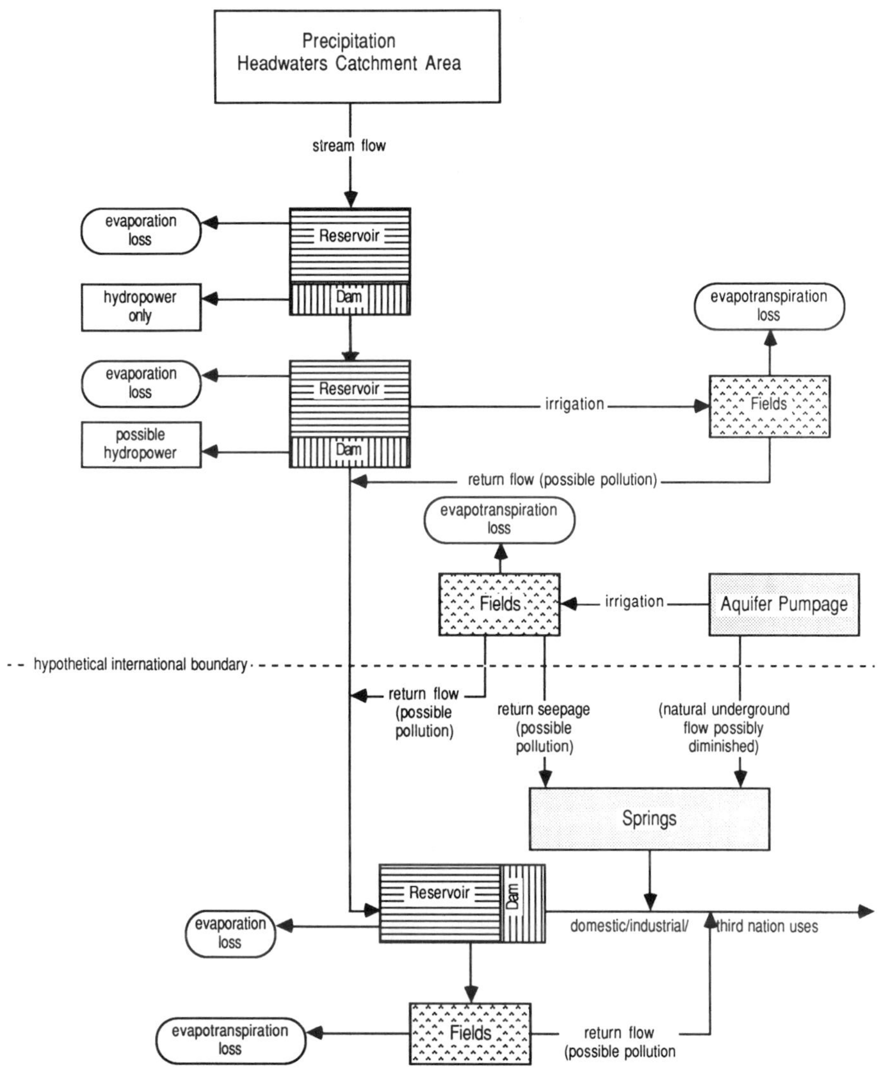

Figure 4.1. Elements of a hypothetical international river use system.

to this general rule. Where spawning runs of fish are concerned, prevention of the breeding stock's progress upstream may reduce fish populations, while the destruction of fingerlings on their way downstream passing through penstocks and turbines can also be a problem. In the case of the Euphrates and Tigris rivers, spawning fish do not present a problem.

A second complication may result from river-borne silt settling in the reservoirs behind dams, whether these dams are intended for hydropower generation or irrigation or both. Excessive quantities of alluvium may fill in reservoirs and reduce their useful lifespan; silt-free waters downstream from such reservoirs may have increased erosive power with subsequent channel changes and/or the undermining of man-made structures. In the case of the Nile, water-borne silt had, before the High Dam, also restored fertility to flooded fields, but this was not true downstream on the Euphrates in Syria. This subject, vis-a-vis Iraq, is not considered here.

Of the six listed uses, irrigation is the most demanding and potentially destructive. For example, it has been estimated that in Egypt agricultural water use represents 92.5 percent of all water extracted from the Nile (interview with John Alan, 1984). A further concern where irrigation is a factor is the quality of the water returned to the main stream after passing through the fields. Heavy loads of fertilizers, insecticides, herbicides, and dissolved natural salts can make water unpalatable and even unusable for further irrigation. (This topic is treated more fully in Chapter 10, "Sedimentation and Water Quality.") Pollution from domestic and industrial use can also be a problem, although the low level of such use in the GAP area (see Chapter 3, Industry and Potable Water for Domestic Use) diminishes this as an issue. As mentioned elsewhere, navigation is essentially out of the question on the upstream portions of the Euphrates and Tigris rivers, and fishing is of little consequence.

Another source of water, which may be independent of stream flow but which may play an important part in determining the quantity and quality of available water, is pumping from underground reservoirs or aquifers. In the case of the Euphrates-Tigris river basin, the aquifers which supply the Khabur River in northern Syria are for the most part located north of the border in Turkey. As will be shown in Chapter 9, although the conventional view is that the Khabur and its tributaries provide up to 12 percent of the flow of the Euphrates, the sources of these streams, and also those of the Balikh farther west, are springs rising just inside Syria south of the Turkish border. These springs receive most of their water, in turn, from large pervious catchments to the north in Turkey which are areas of higher rainfall. Prior to new development plans in Turkey, these springs and the streams dependent upon them represented an inviolate Syrian resource. Now, however, the Turks plan to pump large quantities of water from these aquifers in their own territory. The issue of underground water rights is extremely complicated, and certainly Turkey as well as Syria should benefit from this resource. Nevertheless, this is another possible source of international conflict unless it is understood and resolved by negotiation.

Furthermore, while depletion of underground waters is a major consideration, there is also the question of return flow to streambeds and to underground conduits or aquifers. If the quality of the water running off the fields and/or seepage back into the aquifers is significantly lowered, this can seriously affect downstream use. If any group is to suffer from this phenomenon along the Euphrates, it will be the Iraqis who are farthest downstream.

It should also be noted that return flow from irrigated fields will be reduced in quantity because of inefficient use of the delivery system (canals, storage depots, pumping stations) and through similar inefficient use and application of water on the farms themselves. System efficiency in Turkey and Syria is discussed in Appendix A. Another source of water depletion is the amount used by plants (crops and weeds) to maintain their metabolisms (transpiration), and also the water evaporated from surfaces (soil, stalks, leaves, etc.). These two losses to the atmosphere are subsumed under the term evapotranspiration. Thus a large part of the water removed from rivers and reservoirs for irrigation will not find its way back into the river. Return flow as such has been estimated for the purposes of this study to be approximately 35 percent of the water withdrawn from the system (see Appendix A). (For a more complete discussion of the characteristics of Middle East river system use, see Kolars [1988].)

River Systems—An Overview

Let us now take an overview of the systems which have briefly been described. Figure 4.1 illustrates elements Middle Eastern rivers have in common. The diagram is simplified so that it can be applied to numerous examples throughout the region. Stream flow begins with natural precipitation at the headwaters of country number one. Water may be impounded for the generation of hydropower, with some possible loss through evaporation off reservoir surfaces. Water then continues downstream to the next reservoir, which not only is used to generate electricity but also serves to irrigate fields. Evaporation losses occur from the surface of the second reservoir; losses also occur from fields through evapotranspiration and through system inefficiencies (leakage from ditches, evaporation from open channels, etc.). Return flows may or may not be unacceptably polluted.

Farther downstream, pumpage from independent aquifers irrigates additional fields and provides some return flow, which may increase downstream quantities but may also increase their salinity. Losses also occur through local evapotranspiration. Return seepage from fields may restore some portion of the water removed through pumping but may also pollute spring waters. Excessive pumping may diminish spring flow "downstream" on the aquifer and even across the international frontier. (Lag time because of storage capacity of the aquifer as well as difficulty of observation may make cause and effect difficult to establish in this case.) In country number two similar patterns are repeated, all of which can have implications for countries farther downstream. At all points along the river, changes in the amounts and quality of water may affect domestic and industrial use. These situations can and do occur in numerous permutations and combinations. At the same time, it should be kept in mind that aridity and water need increase as you move from the headwaters downstream, just as, conversely, precipitation diminishes in the same direction.

With the above description of river use in mind, it is time to consider the approach used in the remainder of this study.

Organization of the Analysis

While the intent of this study is to analyze in as much detail as possible the impact of Turkish development plans on the Euphrates River, the approach to this topic begins with a discussion of the actual amounts of water involved. This, in turn, first necessitates looking closely at Syrian data and development plans before turning to a similar analysis for Turkey. Thus the discussion and tables in Chapter 10 will present the Turkish case and articulate its analytical results with those downstream in Syria. This analysis ends at the Syrian/Iraqi border (with reference to Hit, Iraq) because of the special conditions prevailing in the latter country.

The following chapters will examine the use of Euphrates River waters in logical, though not necessarily geographical, sequence. Chapter 5, "The Annual Discharge of the Euphrates River: Turkey into Syria and Syria into Iraq," provides a discussion critical to any planning and/or negotiations regarding the amount of water available to be used by each of the three riparian states involved.

A discussion of "The Euphrates System in Syria" and its average discharge increments to the tributaries follows in Chapter 6. The defining of such shares of river flow is necessary before an analysis of Syrian use—actual and projected—can be attempted.

A further step must precede such an analysis. This refers to the actual amounts of water that must be applied to each unit of developed land in order to meet irrigation requirements based on climatological, soil, and crop conditions. Chapter 7, "Water Use per Hectare and Anticipated River Depletion," undertakes this task in terms of both Syrian and Turkish usages.

Once the amount of water necessary for successful irrigated farming has been determined, it is necessary to learn the actual amounts of land currently irrigated and subsequently scheduled for irrigation. Chapter 8, "Irrigated Agriculture in the Syrian Euphrates Drainage Basin," considers the numerous reports associated with this topic and suggests figures compatible with available data. A similar presentation will be made in the summary section for Turkish irrigated lands although these have already been referred to in the earlier chapters of this study.

One further area of investigation must be considered before a final summary analysis of the Euphrates is given. This is the nature of the Khabur River and its tributaries in the Jezirah of northeast Syria. Chapter 9 points out that the flow of the Khabur, upon which Syria places so much emphasis, is in fact largely derived from and controlled by catchment areas inside Turkey. "The Khabur River and Its Tributaries" spells this out in detail sufficient to make this an issue of concern to planners and politicians.

Chapter 10, "Static and Dynamic Views of the Euphrates River System," will it is hoped speak for itself. Chapter 11, "The Years Ahead," takes a look at the probability of future developments as well as attempting to estimate the timing of such events. Ancillary issues such as environmental impacts and ethnic issues are touched upon, as well as any last minute developments.

5

Average Annual Discharge of the Euphrates River: Turkey into Syria, Syria into Iraq

If and when tripartite negotiations take place concerning the use of Euphrates River waters, much will depend upon a clear understanding of the quantity available at any given time to be shared among the riparian users. The first such measure concerns the average annual discharge of the river. This is no simple matter to determine, for it seems that every report and evaluation quotes a different set of figures. Moreover, Turkish reports disagree with other analyses for the same gauging stations, as do those for Syria and Iraq.

Table 5.1 lists six stations along the river from Birecik, near the border in Turkey, to Hit in Iraq. The 11 sources of information list 17 values, excluding one estimated "natural flow" (see Table 5.1, footnote 3), none of which agree and few of which offer consistent data. Possibly other references could be found listing still more flow or discharge data, but those would only add to the confusion. The only new materials which could clarify the situation would be complete flow records from at least one major station in each country for the same long span of years, measured in the same way in each case. It is unlikely that such a data trove will become available. On the other hand, some sense can be made of all this if the accompanying tables and graphs are carefully examined along with the text that follows.

Table 5.1. Discharge of the Euphrates from Birecik, Turkey, to Hit, Iraq

Location	Flow in m³/s min	max	aver	Flow in Mm³/yr	Time Period	Data Source
Birecik	484	1356	856	26,990	1937–63	Clawson
Birecik	—	—	—	30,970	?	GAP
[Average, N=2]				28,980	?	
Karkamis	—	—	—	31,380	?	GAP
Syrian/Turkish Border	—	—	—	22,100	7 yrs	"last 7 yrs." USAID, II-7
Yusuf Pasha	—	—	—	26,050	1955–66	UNESCO, noted in Beaumont, 40
Tabqa	—	—	—	26,000	1973	Samman
Tabqa	—	—	913	28,790[a]	21 yrs?	al-Hadithi
Tabqa	—	—	810	25,543[a]	14 yrs	See Table 6.5
Tabqa	—	—	735	23,180[a]	?	Shchukin
Tabqa	—	—	—	26,200	?	Wirth/Bourgey, 343
Tabqa	—	—	—	23,950	10 yrs	USAID
[Average]				27,230	31 yrs	al-Hadithi & USAID
Hit	—	—	927[b]	29,240	?	Clawson, 205
Hit	—	—	—	33,690[c]	?	Clawson, 205
Hit	535	1378	934	29,450	1937–63	Clawson[d]
Hit	—	—	902	28,400[a]	1924–73	al-Hadithi[e]
Hit	—	—	931	29,360[a]	1924–78	GOI[f]
Hit	—	—	1009	31,820[a]	1940–69	Ubell, 4
[Unweighted average, N=4]						
Hit	—	—	853	26,900	1978	al-Hadithi[g]

Source: Clawson et al. (1971); GAP (1980); USAID (1980); Beaumont (1978); Samman (1980); al-Hadithi (1978); SAR (1980); Shchukin (1967); Bourgey (1979); Ubell (1971).

[a] Computed from average flow.
[b] Measured flow.
[c] Estimated "natural flow."
[d] By inspection, Clawson (1971, Table B-10) is a subset of al-Hadithi (1978, Appendix E).
[e] Average of flow values given in al-Hadithi (1978, Appendix E).
[f] Government of Iraq, Ministry of Irrigation, as cited in al-Hadithi (1978, 47 and 52, Table 3.)
[g] al-Hadithi (1978, 114).

Figure 5.1 displays the information given in Table 5.1, with upstream data on the left and downstream data on the right. The points indicated in every case are identified by the source from which that value is derived. A discussion of these sources of data helps to identify what may be the most accurate picture of average yearly discharge. The lines joining the upper row of values, as well as the ones joining the lower values, do not imply natural sequences of flow, but rather are meant to indicate reasonable upper and lower limits on such data.

Discharge in Turkey

Birecik, Turkey, shows two divergent values for discharge at that point. The greater value is drawn from the *Southeast Anatolia Development Project Report* (GAP 1980). (These comments also apply to the single value for Karkamis, downstream from Birecik almost to the Syrian border.) In this case, neither the number of years nor the specific years involved are mentioned in the original report. The lower value for Birecik is drawn from Clawson et al. (1971), who in turn cite Hathaway et al. (1965) and their 1965 International Bank for Reconstruction and Development report on the Keban Dam (Table 5.2). This is based on the 27 years from 1937 through 1963, with partial data for 1964. Evidence discussed below suggests that the GAP figures are for a shorter and more recent period of time. The accuracy of the Clawson data might be questioned pro forma, but at least the time span is known. Similar data from Clawson for Hit, Iraq (Table 5.3) can be shown by inspection to be a subset of the data provided al-Hadithi (1978) by the Iraqi Ministry of Irrigation (Table 5.4). As will be seen, these data seem to be consistent and usable. By inference, the Clawson data for Birecik should be reasonably reliable. GAP data are probably accurate for the years they represent, but much depends upon the number of years and the time span chosen when considering a river with as irregular a regime as that of the Euphrates.

The slightly higher figure given by GAP (1980) for Karkamis, downstream from Birecik, is consistent with the former's geographical situation. As mentioned in Chapter 6 regarding the Eu-

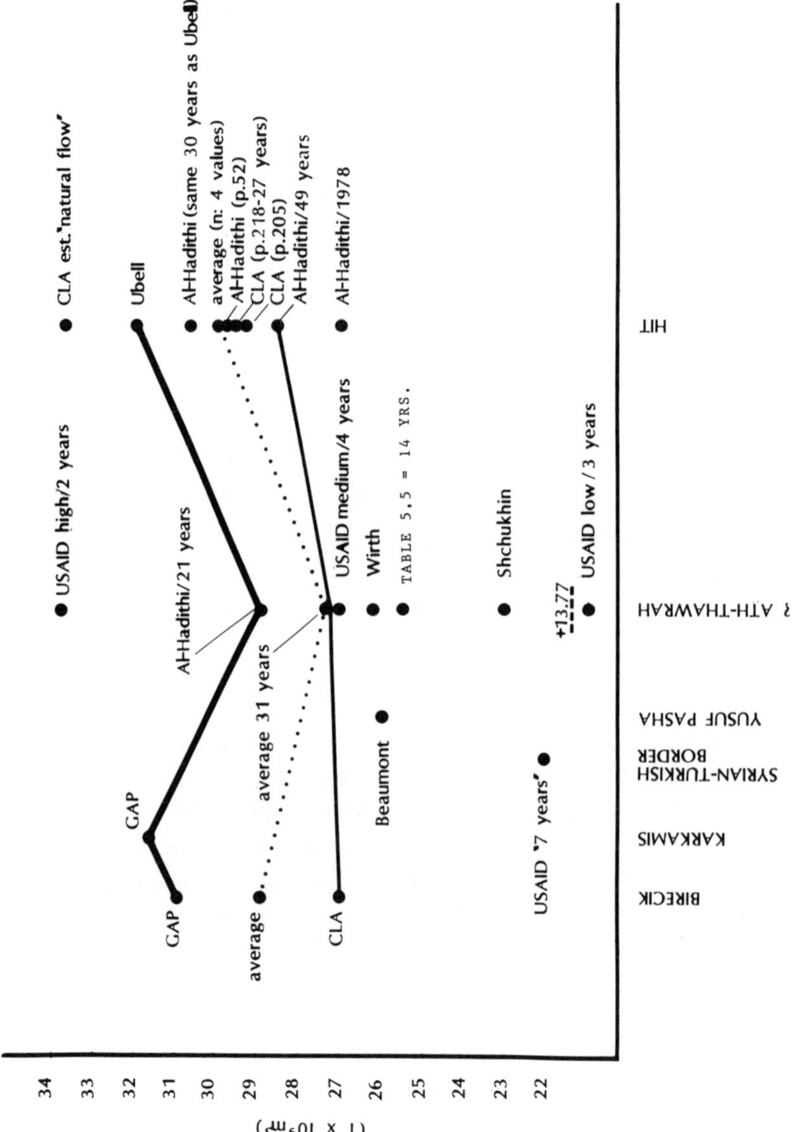

Figure 5.1. Reported average yearly discharge of the Euphrates.

phrates in Syria, tributary flow from the Nizip and other small streams in Turkey should account for this increase. Nevertheless, both these GAP values appear unusually large.

Flow in Syria

GAP data (1980) are in sharp contrast with the next two values on this chart. The USAID report on Syrian agriculture (USAID 1980) quotes an overall flow for the Euphrates of 27,000 Mm^3/ yr, but qualifies its statement by adding, "The flow of the Euphrates the last seven years has averaged substantially less, however, about 22.1 Mm^3; measurements at the Syrian-Turkish border." The report also lists the flow for the years 1967 and 1970–1977 (Table 5.5). Flows for 1978, 1979, 1980, 1982, and 1984 are also available (SCBS 1979, 1980, 1981, 1983, 1985). The average flow for the years 1973–1979 is 747 m^3/s or 23,566 Mm^3/yr, somewhat more than the quoted 22,100 Mm^3 shown on the graph, but the question remains that the location where these data were taken is unspecified and may be downstream beyond the confluence with the Sajur, thus possibly accounting for the increased value.

Inspection of these data (Table 5.5 and Figure 5.2) shows wide fluctuations ranging from 12,800 Mm^3/yr to 32,860 Mm^3/ yr in the space of 36 months. Values for the earlier years in this series come close to the 27,000 Mm^3 quoted by USAID (1980) and seem consistent with Clawson (1971) data if an additional downstream increment were taken into consideration. Values for 1973–1975 appear anomalous at first and far too low. It was, however, in the winter of 1973–1974 that the Keban and Lake Assad reservoirs began to be filled. On the other hand, inspection of flow data for other rivers and streams in Syria (Table 5.6 and Figure 5.2) for the same time period show a significant diminution of discharge throughout the country, outside as well as inside the Euphrates drainage basin. This period of low discharge on the Euphrates cannot be explained through reservoir filling alone and is undoubtedly climatic in origin. Had this information been known at the time, the near

Table 5.2. Discharge of Euphrates River at Birecik, Turkey, 1937–1964
(average flow in m³/s)

Year	Jan	Feb	Mar	Apr	May	Jun	Jul	Aug	Sept	Oct	Nov	Dec	Annual Average
1937	496	667	4328	2644	1753	870	522	370	336	369	573	798	894
1938	680	647	873	3257	2686	1084	638	398	347	345	448	557	997
1939	558	620	1061	2382	2126	717	482	349	445	316	374	547	831
1940	848	947	985	4081	2742	1190	677	391	326	458	503	837	1165
1941	725	1362	2669	3188	2434	750	483	344	304	330	379	473	1120
1942	504	581	1038	3105	2886	994	448	323	292	360	898	957	1032
1943	603	544	696	2853	2548	853	421	308	278	311	340	514	856
1944	468	699	2019	2336	3339	1129	563	387	351	361	500	518	1056
1945	560	53	697	1911	1871	868	361	271	241	238	268	472	691
1946	417	422	947	2337	2868	1246	571	397	307	617	416	499	920
1947	580	620	1697	1839	1028	570	328	252	227	232	620	448	703
1948	442	627	583	3216	3376	1566	583	356	297	297	320	426	1007
1949	381	420	644	1773	2234	737	349	278	250	254	258	371	662
1950	359	423	824	2164	2305	727	399	313	282	405	358	472	753
1951	505	501	1203	1983	1449	676	355	271	279	408	432	534	716
1952	440	838	989	3438	2418	957	458	315	289	285	299	456	932
1953	462	643	743	2927	2658	1207	538	332	285	292	365	419	906
1954	430	476	1184	3382	2972	1234	609	345	302	304	340	561	1012
1955	505	545	802	1367	1424	560	238	279	249	252	277	465	588

1956	440	509	689	2720	2300	1080	488	335	304	309	316	437	827
1957	420	577	1641	1766	2336	1027	455	303	270	271	301	449	818
1958	450	506	1059	1943	1385	740	347	268	240	240	259	419	655
1959	369	315	754	1701	1351	718	302	249	239	261	301	327	574
1960	640	529	1055	3005	2098	740	400	297	272	252	297	326	826
1961	363	495	534	1409	1038	426	219	174	156	177	293	525	484
1962	434	805	1443	1710	1240	637	359	254	229	239	293	659	692
1963	1008	1118	1025	3291	4115	2321	951	508	410	510	507	512	1356
1964	369	540	1837	2528	1781	911	400	267	249				

Mean in

m³/s	516	625	1108	2509	2241	948	466	319	288	324	393	521	
Mm³	1380	1520	2970	6500	2460	1250	850	750	870	1010	1400		

Source: Clawson (1971, 217, Table B-9).

Table 5.3. Discharge of Euphrates River at Hit, Iraq, 1937–1964
(average flow in m³/s)

Year	Jan	Feb	Mar	Apr	May	Jun	Jul	Aug	Sept	Oct	Nov	Dec	Annual Average
1937	525	620	1070	2080	1800	1090	558	343	275	291	679	1090	862
1938	1130	971	966	2220	3200	1450	778	461	355	359	491	528	1076
1939	731	761	1120	2000	2530	1230	685	452	375	352	395	586	935
1940	1010	1080	1270	3060	2950	1330	700	418	343	407	708	908	1182
1941	922	1300	2700	2700	2420	1040	549	303	321	341	417	390	1117
1942	603	821	1220	2640	3030	1190	451	281	238	329	919	1210	1078
1943	1220	979	990	2350	2990	1190	573	376	309	330	483	485	1023
1944	666	754	1650	2250	3210	1400	622	394	359	379	634	513	1069
1945	904	726	847	1670	2120	1420	630	358	290	304	371	574	851
1946	580	656	1130	2160	3100	1660	765	463	376	591	612	473	1047
1947	870	900	1560	2080	1140	745	449	301	261	281	549	575	809
1948	554	1160	919	2560	3560	1950	749	408	349	356	368	495	1119
1949	407	542	585	1670	2200	1120	472	319	273	283	311	355	711
1950	448	354	1010	1970	2520	1130	494	311	264	315	401	373	799
1951	554	503	764	1870	1580	836	371	246	226	399	471	576	700
1952	451	1270	1140	2940	2350	1160	558	334	281	308	343	416	963
1953	537	1030	1310	3010	3110	1660	712	397	342	359	483	478	1119
1954	644	890	1630	3820	3380	1670	761	423	336	373	508	617	1254
1955	1090	706	899	1410	1720	777	340	228	228	284	318	521	710

1956	751	819	988	1750	2730	1230	558	314	269	328	370	402	876
1957	362	441	1580	1640	2690	1520	588	293	238	300	417	649	893
1958	679	644	1080	1820	1560	1140	414	219	196	307	373	495	744
1959	502	474	664	1672	1513	1029	364	209	194	290	390	360	638
1960	881	607	1298	2684	2766	1177	522	303	253	355	418	412	973
1961	494	571	466	138	1209	475	197	94	99	191	377	905	535
1962	694	979	1338	1835	1454	883	298	153	317	248	297	491	749
1963	851	1300	1365	2585	4368	2819	931	422	311	451	630	505	1378
1964	398	468	1218	2621	1597	1075	373	168	172				
Mean in													
m^3/s	695	797	1171	2229	2457	1264	552	321	280	337	468	576	
Mm^3	1860	1940	3140	5770	6540	3270	1480	860	730	900	1210	1540	

Source: Clawson (1971, 218, Table B-10).

Table 5.4. Mean Monthly and Mean Annual Discharges of Euphrates River at Hit, Iraq, 1924–1973
(flow in m³/s)

Years	Oct	Nov	Dec	Jan	Feb	Mar	Apr	May	Jun	Jul	Aug	Sep	Annual
1924–25	261	299	713	369	309	650	1014	1117	829	555	255	212	549
1925–26	229	288	488	714	737	1086	2296	2344	1416	646	357	270	906
1926–27	256	374	375	344	356	507	1284	1506	729	354	269	222	547
1927–28	231	266	264	291	388	543	1951	1718	651	345	241	230	593
1928–29	227	277	587	507	710	885	2198	3358	1758	712	460	337	1001
1929–30	346	334	435	335	342	344	481	533	249	268	202	283	338
1930–31	225	265	367	568	605	798	1794	1898	1367	635	355	278	763
1931–32	303	308	350	343	350	750	1270	1620	834	375	242	213	580
1932–33	232	250	270	275	313	481	501	1600	1110	443	236	215	495
1933–34	209	231	317	398	448	687	1530	1270	930	413	306	242	582
1934–35	236	341	268	681	885	1260	2560	2330	939	528	397	350	889
1935–36	356	747	1310	879	1140	1290	2250	2530	1630	811	519	331	1140
1936–37	321	382	757	525	620	1070	2080	1800	1090	558	343	275	819
1937–38	291	679	1090	1130	971	966	2220	3200	1450	778	461	355	1130
1938–39	359	491	528	731	761	1120	2000	2530	1230	685	452	375	939
1939–40	352	395	586	1010	1080	1260	3060	2950	1330	700	418	343	1120
1940–41	407	708	908	922	1300	2700	2700	2420	1040	549	303	321	1190
1941–42	341	417	390	603	821	1200	2640	3030	1190	451	281	238	969
1942–43	329	919	1210	1220	979	990	2350	2990	1190	573	376	309	1120
1943–44	330	483	485	666	754	1650	2250	3210	1400	622	394	359	1050

1944–45	379	634	513	904	726	847	1670	2120	1420	630	358	290	874
1945–46	304	371	574	580	656	1130	2160	3100	1660	765	463	376	1015
1946–47	591	612	473	870	900	1560	2080	1140	745	449	301	261	830
1947–48	281	549	575	554	1160	919	2560	3560	1950	749	408	349	1130
1948–49	356	368	495	407	542	585	1670	2200	1120	472	319	273	734
1949–50	283	311	355	448	354	1010	1970	2520	1130	494	311	264	789
1950–51	315	401	373	554	503	764	1870	1580	836	371	246	226	670
1951–52	399	471	576	451	1270	1140	2940	2350	1160	558	334	281	991
1952–53	308	343	416	537	1030	1310	3010	3110	1660	712	397	341	1100
1953–54	359	483	478	644	890	1630	3820	3380	1670	761	423	336	1240
1954–55	373	508	617	1090	706	899	1410	1720	777	340	228	228	742
1955–56	284	318	521	751	819	988	1750	2730	1230	558	314	269	877
1956–57	328	370	402	362	441	1580	1640	2690	1520	588	293	238	874
1957–58	300	417	649	679	644	1080	1820	1560	1140	414	219	196	760
1958–59	307	373	495	502	474	664	1672	1513	1029	364	209	194	650
1959–60	290	390	360	881	607	1298	2684	2766	1177	522	303	253	961
1960–61	355	418	412	494	571	466	1338	1209	475	197	94	99	510
1961–62	191	377	905	694	979	1338	1835	1454	883	298	153	317	785
1962–63	248	297	491	851	1300	1365	2585	4368	2819	931	422	311	1332
1963–64	451	630	505	398	468	1218	2621	1597	1075	373	168	172	806

continued

Table 5.4. (continued)

Years	Oct	Nov	Dec	Jan	Feb	Mar	Apr	May	Jun	Jul	Aug	Sep	Annual
1964–65	291	333	603	467	841	1210	2120	2245	1210	483	256	218	855
1965–66	408	513	652	1194	2118	1514	2241	2649	1451	583	325	304	1155
1966–67	534	634	849	988	968	1281	2787	4920	2199	999	491	408	1424
1967–68	644	1091	1259	1319	1263	2376	3794	4185	2271	956	495	467	1677
1968–69	557	725	1586	2448	1697	2732	4589	5460	2307	968	535	488	2011
1969–70	526	567	569	641	761	1122	1786	1114	602	249	147	165	595
1970–71	297	332	582	386	350	830	2522	1717	1027	402	249	254	746
1971–72	376	413	516	406	401	700	1495	2319	1367	511	229	243	809
1972–73	330	461	414	317	337	595	1130	1280	608	192	81	89	526

Source: al-Hadithi (1978, 225–27, Table 1).

Table 5.5. Discharge of Euphrates in Syria[a]
(data for available years)

Year	m^3/s	Mm^3/yr	Average
1967	830.0	26,170	
1970	835.0	26,330[b]	
1971	835.0	26,330[b]	
			(n=4)
1972	835.0	26,330[b]	26,290 Mm^3
1973	476.0	15,010	
1974	406.0	12,800	
			(n=3)
1975	428.0	13,500	13,770 Mm^3
1976	1100.0	34,690	
			(n=2)
1977	1042.0	32,860	33,780 Mm^3
1978 [c]	971.0	30,621	
1979 [c]	808.0	25,480	
		22,100[d]	7 yr average
1980[c]	1013.0	31,944	
	998.0	25,172	12 yr average
1982[c]	1063.0	33,522	
1984[c]	698.0	22,012	—
	810.0	25,543	14 yr average

Source: USAID (1980, II-32–34, Tables 3 and 4).

[a] Location of the gauging stations unidentified in original source.
[b] Values for these 3 years are identical in USAID.
[c] Source: SCBS (1979; 1980; 1981; 1983; 1985).
[d] USAID (1980, II-7): "The long-term annual quantity of water in streams and rivers in Syria is quoted at 32 billion m^3 in the Euphrates river. . . . the flow of the Euphrates the last seven years has averaged substantially less, however, about 22.1 billion m^3; measurements at the Syrian-Turkish border."

confrontation between Syria and Iraq over the diminished river flow might have been avoided.

The next value is given by Beaumont (1978) for Yusuf Pasha near the head of Lake Assad, upstream from the Tabqa Dam in Syria. This is an average for 17 years from 1950 through 1966, a period of relatively low water in the entire system (Figure 7.7). It

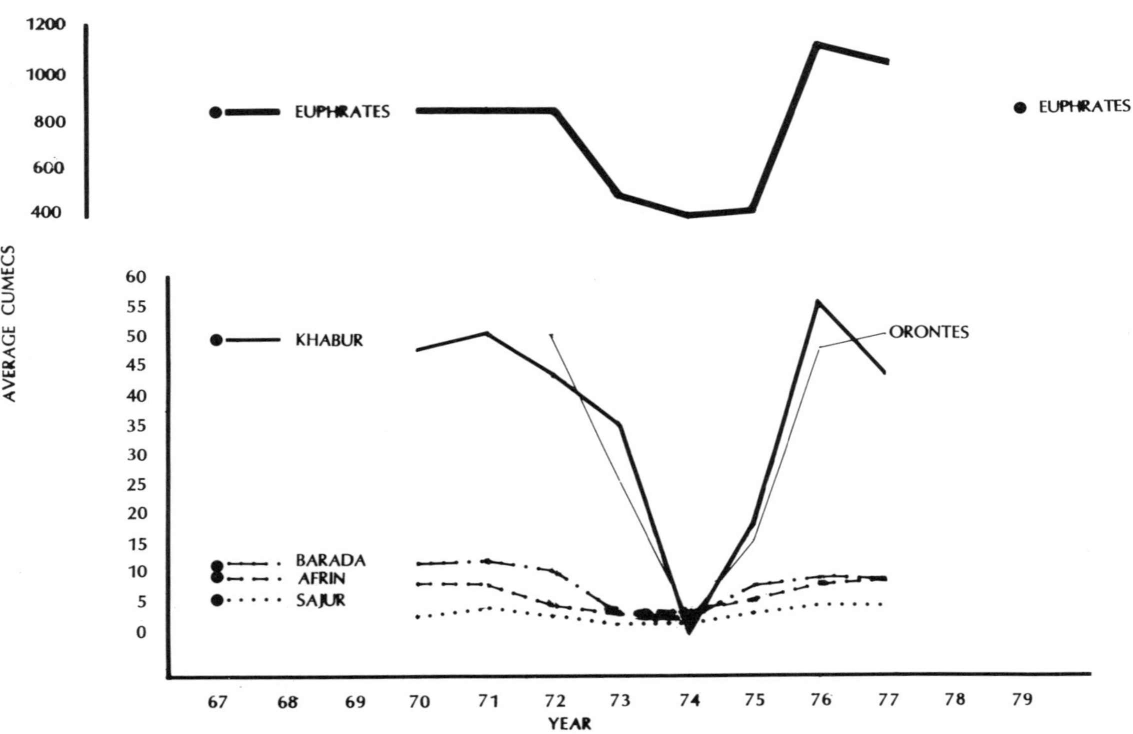

Figure 5.2. Annual discharge of selected Syrian rivers. Source: USAID (1980, Tables 3 and 4); SAR (1980).

Table 5.6. Discharge Data for Selected Rivers in Syria (in m³/s)

Year	Euphrates	Khabur[a]	Barada	Afrin[a]	Sajur	Orontes[a]
1967	830.0	48.0	8.8	8.0	4.0	
1970	835.0[b]	48.0	11.0	8.0	3.0	
1971	835.0[b]	50.0	11.0	8.0	4.0	
1972	835.0[b]	43.0	10.0	4.0	3.0	50.0
1973	476.0	35.0	2.4	3.0	1.5	25.0
1974	406.0		3.0	0.7	0.4	
1975	428.0	18.0	4.2	5.3	3.9	16.9
1976	1100.0	56.4	8.7	9.3	4.8	49.7
1977	1042.0	43.0	8.8	8.0	4.2	51.1
1978[c]	971.0		10.4	9.7	4.2	60.1
1979[c]	808.0		3.6	9.7	1.9	34.3

Source: USAID (1980, II-32–34, Tables 3 and 4).

[a] And tributaries.
[b] 3 consecutive years with the same figure in the original source.
[c] Source: SCBS (1979; 1980).

should be kept in mind that the close correlation between Hit data and Birecik data shown in Figure 7.3 permits some interpretation of points in between the two stations.

The 14-year average (Table 5.5) cited above has been placed on Figure 5.1 at Tabqa, where it still appears as a somewhat low value for the site. Shchukin (1967) gives the lowest value without reference to the time span or dates covered. Indeed, it is so low that it suggests that he may be citing a single year's discharge. The value quoted by Samman (1980) in Table 5.1 is inconsistent with USAID data and suggests that he has cited a wrong year (possibly 1972). Therefore, his datum is not shown on the graph and is mentioned here only to illustrate the difficulties surrounding these evaluations. Low, medium, and high USAID (1980) averages are also shown for comparison.

It should be noted that the average value for 14 years shown in Table 5.5 is consistent with the low value given upstream by Beaumont (1978). Little is known regarding Wirth's value, discussed in Bourgey (1979). It merely reinforces the idea that long-run average flow rates should have lower values than that quoted by al-Hadithi (1978), whose higher value is for 21 unidentified years, presumably in a consecutive sequence. The top USAID (1980) datum is as unusually high as Shchukin's (1967) is low. This figure covers a short time span (only two years) and represents an infrequent period of flooding.

Flow in Iraq

There are eight values available for Hit, Iraq, and it is these which allow some estimation and evaluation of the correctness of the various data given in Table 5.1. The lowest al-Hadithi (1978) datum is for a single year and is consistent with the lower range of river flow. The second and larger value (moving up the column) is for 49 years from 1924–1925 through 1972–1973 (Table 5.1). Al-Hadithi cites the Iraqi Ministry of Irrigation as his source for these data. Inspection shows that Clawson (1971) uses a subset of these data, but since Clawson's publication date precedes that of al-Hadithi, the two authors must draw their data from a third common independent source, undoubtedly that cited by al-Hadithi.

The data clustered about the average value (n = 4; see Table 5.1) in the column include one based by Clawson upon a shorter run of years. The same is true for al-Hadithi's value given in that grouping. The next highest al-Hadithi value is for 30 years, but this is lower than that given by Ubell (1971) for the same period—1940–1969. This inconsistency persists when the data provided by both are compared by decades as well as for the entire 30-year period. (Ubell presents his data in increments of ten-year averages.) No reason is given for the discrepancies shown by Table 5.7, and it is unlikely that either writer knew of their existence.

While various explanations suggest themselves, it would serve little purpose to pursue them at this point. Rather, the al-Hadithi data (Table 5.4) provide us with a fairly long and consistent view of river flow. (This includes by extension the subset used by Clawson as given in Table 5.3, but the longer time span is preferable.) Ubell's data would seem consistent with a higher range of values and, as such, may be misleading. The present

Table 5.7. Differences in Data Regarding Discharge of
Euphrates at Hit, Iraq, 1940–1969
(in 10-yr averages)

Years	Ubell		al-Hadithi	
	m^3/s	Mm^3/yr	M^3/s	Mm^3/yr
1940–49	1004	31,700	891	28,100
1950–59	871	27,500	965	30,400
1960–69	1151	36,300	1056	33,300
1940–69	1009	31,800	971	30,600

Source: Ubell (1971, 4); al-Hadithi (1978, Table E-I).

analysis prefers to adopt the more conservative view of the situation. The very high value of 33,700 Mm^3/yr for "natural flow" at Hit—the last item on Figure 5.1—raises the question of the difference between "observed or measured flow" and "natural flow." The latter term refers to stream volume before any human withdrawals have occurred and, as such, must be estimated rather than measured. This last high value is probably a correct estimate (as demonstrated in Chapter 10). Other values discussed here are often apparently "observed or measured" and do not take withdrawals into account.

The Average Regime

What then can be said about the quantity of water in the Euphrates at Birecik, Tabqa, and Hit? It appears that the sequence of data used by Clawson (1971) at Birecik is better than the higher figures shown by GAP (1980). How were the GAP data derived? Table 5.8 suggests an explanation. This table shows the data used by al-Hadithi (1978) for Hit aggregated in 10, 20, 30, 40, and 49-year periods. Note how river flow can vary from one ten-year period to the next (left-hand column). Also note how increasing aggregations can change and/or obscure high and low periods of flow. While these data represent conditions at Hit, the figures in parentheses are approximations of matching flow leaving Turkey. These latter values were derived by reducing the Hit figures by 6.6 per-

Table 5.8. Cumulative Mean Annual Discharges at Hit, Iraq (in multiple decade intervals)

Decade Ending	10-year Average	20-year Average	30-year Average	40-year Average	49-year Average
1933–34	20,040 (18,800)				
1943–44	32,690 (30,670)	26,370 (24,740)			
1953–54	29,560 (27,730)	31,130 (29,200)	27,430 (25,730)		
1963–64	26,170 (24,550)	27,870 (26,140)	29,470 (27,650)	27,150 (25,470)	
1972–73	33,440 (31,370)	29,810 (27,960)	29,720 (27,880)	30,470 (28,580)	28,380 (26,620)[a]

Source: al-Hadithi (1978, Table E-I).

Note: Figures in parentheses represent the estimated flow at Birecik based on an average contribution from Syrian tributaries of the Euphrates of 6.6 percent, i.e., the flow at Hit equals 106.6% of the flow at Birecik.

[a] This figure differs slightly from the true 49-year average for 1924–25/1972–73 (28,440 Mm^3/yr) because of the double counting of 1963–64 and rounding errors.

cent, the average amount shown to enter the system from the Balikh and Khabur rivers in Syria. Without claiming overmuch for evidence such as this, the correspondence between the derived flow for the period 1963–1964 through 1972–1973 and the data given by GAP for Karkamis should be noted. One may ask if the unspecified time period upon which GAP data are based perhaps corresponds to this decade of river flow.

Table 5.9 further illustrates the variability and complexity of discharge. The four years of greatest flow (1965–1966 through 1968–1969) have the phenomenal average of nearly 50,000 Mm^3/ yr. The smallest flow for a consecutive 4-year period averaged about 17,000 Mm^3/yr. Within those 4-year periods, the single largest annual flow equaled 63,000 Mm^3 and the least 10,700 Mm^3. Figure 6.7 shows the 49-year series at Hit. The flood of 1969 catches the eye and dominates. One may ask in the words of P. J. Weatherhead (1986) "How unusual are unusual events?" In his

Table 5.9. Ten-Year Average Discharges of Euphrates at Hit, Iraq, 1924–1973

	Years (Oct–Sep)	Flow in m^3/s	Flow in Mm^3
	24/25–33/34	635.4	20,040
	34/35–43/44	1036.6	32,690
	44/45–53/54	937.3	29,560
	54/55–63/64	829.7	26,170
	63/64[a]–72/73	1060.4	33,440
49 years	24/25–72/73	901.8	28,440
Four largest consecutive years	65/66–68/69	1566.75	49,410
Four smallest consecutive years	29/30–32/33	544.0	17,160
Largest year	68/69	2011.0	63,420
Smallest year	29/30	338.0	10,660

Source: Iraqi Ministry of Irrigation, given in al-Hadithi (1978, Table E-I. Breakdown computations by Kolars.

[a] The year 1963–64, which was used in the preceding decade's calculations, is included here to give a 10-year average.

review of unusual events and their impact on ecological and biological systems Weatherhead concludes, "We tend to overestimate the importance of some unusual events when we lack the perspective provided by a longer study." In this case, 49 years of data do not seem long enough to provide an objective perspective. Again, as the statistician M. J. Moroney says, "I dislike time series and index number men. The plain truth is that we can never—except by an act of great faith—say that an existing trend will be maintained even for a short time ahead" (Moroney 1951, 372). If one were to fit trend lines to the data shown in Figure 6.7, there would be some upward slope from 1930 to 1969. But the period from 1941 to 1961 would show a downward trend. We are even further blinkered by lacking data for the last 15 years (1974–1988). At least, with 49 years available, the lean years of the thirties tend to balance out the abundant late sixties. Such differences present opportunities for choices based on political points of view—a fact to be remembered.

To continue downstream from Birecik, the slight increase at Karkamis is consistent with the regime of the river, but would a parallel upward value persist at that point if long-run data were available? It seems likely that would be the case. USAID's (1980) average "for the last seven years" reflects the unusually low water from 1973 through 1975. Whether drought or removals account for such a deficit, this seems far too low for long-range planning. Beaumont's (1978) datum for Yusuf Pasha is in a range similar to the lower values shown for Tabqa (Figure 5.1). Should we then reject the high values at Tabqa cited by al-Hadithi (1978)? After all, he says they cover 21 years.

Since only the Sajur contributes to the river between this location and the Turkish border, a slight increase suggested by the lower limit line seems more consistent. It may be that al-Hadithi's choice of 21 years included years with relatively high water levels. If we accept Clawson's (1971) data for Birecik, it is reasonable to expect slightly higher values than those given by the Syrian Central Bureau of Statistics (1980) and USAID's nine-year average for Tabqa. (There is also the possibility that the USAID figures refer to a point at the Syrian-Turkish border, which might account for their being somewhat lower.) It also appears that al-Hadithi's higher value for Tabqa is inconsistent with his other data for Hit. These latter reflect tributary flows downstream from Tabqa—the Balikh and Khabur—and should be greater. This increase between the two stations is shown by the lower limit line. Nevertheless, about 500 Mm^3 of the difference is not accounted for with these data.

Finally, at Hit, al-Hadithi's (1978) data are most complete. Except for the unexplained disagreement between Ubell's data (1971) and his data, the 49-year series al-Hadithi presents is convincing.

Safe Values

The gist of all this is that, if we limit ourselves to a consideration of observed flow or measured flow, then the longer run, lower average values seem safest for talking about future river use. Thus, the data given by Clawson (1971) for Birecik (26,990 Mm^3/yr) and the 49-year record provided by al-Hadithi (1978) for Hit (28,400 Mm^3/yr) represent the best data sets this study can provide (Table 5.10). The data for Tabqa are less certain; a middle range average

Table 5.10. Best Annual Estimate of Euphrates Flow to the
Year 1973
(in Mm^3)

| | Birecik | | Tabqa | | Hit | |
	Year	Amount	Year	Amount	Year	Amount
Maximum recorded[a]	1963	42,700[1]	1976	34,690[2]	68–69	63,420[3]
Minimum recorded[b]	1961	15,260[1]	1974	12,800[2]	29–30	10,660[3]
Average discharge	37–63	26,990[1]	31 yrs	27,230[4]	24–73[c]	28,400[3]
Est. discharge at Karkamis[d]		27,400				

Source: 1=Clawson (1971); 2=USAID (1980); 3=al-Hadithi (1978, 4, Table 5.1).

[a] Higher discharges no doubt occurred at Birecik and Tabqa in 1969. However, no specific and reliable data are available for that year for those stations.

[b] The same may be said of low-water years such as 1929–30 at Hit. Lack of data for Turkey permits only estimates for such periods, as well as those at Tabqa.

[c] Clawson (as a subset of al-Hadithi) provides a value for 1937–63 of 29,450 Mm^3. This provides consistency along the average discharge row, but the longer record for Hit is preferable for many purposes.

[d] This estimate is based upon the materials shown in Table 5.1 and would include the flow of the Nizip tributary. This figure is shown only to remind the reader that additional water can enter the system in Turkey downstream from Birecik. The total, however, is not intended to match this table or the discussion upon which it is based.

value (27,230 Mm^3/yr), although less thoroughly substantiated than the values for Birecik and Hit is consistent with them.

There remains the question of natural flow versus measured flow, for any diplomatic partitioning of the river's waters must be based upon the amounts provided by natural causes. As will be seen in Chapter 10, the amounts of water already being withdrawn in Turkey and Syria have a significant impact on the quantity arriving in Iraq. That complication, however, will be considered in Chapter 9, which is devoted to the Khabur in Syria, as well as in the part of this study analyzing Turkish use of Euphrates waters.

6

The Euphrates System in Syria

Syria, as the second riparian user of the Euphrates River, contributes water to the system and also extracts large amounts of water from it. Small quantities of runoff enter the mainstream from wadis along its right bank, but, with the exception of the flow of the Sajur (Turkish: Sacir), this contribution is ephemeral, unpredictable, and negligible. The discharge of left bank tributaries into the Euphrates is much more significant. The Balikh (Turkish: Culap) and Khabur (Turkish: Habur and/or Circip) support considerable agriculture in Turkey and downstream in Syria;[1] these add from 7 percent according to Kolars (this study) to 12 percent ac-

---

1. It is difficult to estimate exactly how much water is removed from the Khabur in Turkey for agricultural purposes. GAP (1980) reports that 6700 ha are irrigated at the State Produce Farm (Devlet Uretme Çiftligi) and that an "important part" of the water comes from underground sources. It also states that four pumps are used to supply water from the Habur to the "upper elevations." It should also be noted that a reservoir called the "Aride" appears upstream from Ceylanpinar on the Habur on GAP maps, although no reference to it is made in the GAP texts. Finally, GAP reports a total of an additional 2186 ha irrigated in the same region from small ponds or reservoirs.

General descriptions of the State Produce Farm can be found in Urfa Provincial Government (n.d., 207–212). Additional information on agriculture in the Urfa-Harran watershed (i.e., the headwaters of the Culap/Balik) is available in Ayyildiz (1986, 305–328). However, no exact figures are provided from which to estimate exact water extractions.

cording to Beaumont (1978) to the discharge downstream into Iraq. A further issue is raised if the source of the water in these two streams is considered. Because these tributaries have their headwaters in Turkey, an additional 5.8 percent (if Kolars's computations are correct) to 10 percent (if Beaumont's larger figure is true) of the total Euphrates volume may be influenced by Turkish water resource planning. In other words, as much as 98 percent of the flow of the Euphrates may originate in Turkey.

The ambitious plans which Syria entertains for the use of Euphrates waters must be emphasized at this juncture. Of primary interest is the question of exactly how much water actually comes from Syria rather than Turkey. If as Beaumont claims some 12 percent of the total flow is Syrian in origin, then the Khabur and Balikh tributaries are secure sources for irrigation development anticipated in their watersheds by the Syrians. On the other hand, if as much as 98 percent of the water actually originates in and can be controlled by Turkish dams and pumps, the Syrians must carefully evaluate their situation and negotiate these details with the Turks.

The importance of the above statement becomes apparent when the scope of Syrian plans is shown. As of 1982–83, some 313,300 ha of land were scheduled for irrigation. Of these, 143,000 ha were in the study stage, 117,000 ha were being prepared, and 53,000 ha were in production (Metral 1987). The Euphrates Dam (Tabqa) not only provides much of the impounded water for these fields, but in 1979, for example, it produced 60 percent of Syria's electricity (2.5 billion kWh) from eight 103,000 kW turbines, of which six were operating. As Syria's need for electric energy inevitably increases, so too will its need for a secure supply of water in the Euphrates. Any such water coming from within Syria is not only securely under Syrian control, but also the larger Syria's donation to the stream the more bargaining power it will have at the negotiating table.

Thus, the questions that need to be asked at this point are: How much water do the Euphrates and its tributaries discharge into Syria from Turkey and from Syria into Iraq? What are the sources of that water? What demands (extractions, polluting return flows) are currently placed on the stream? And what can

Table 6.1. Euphrates River Discharge from Birecik, Turkey, to Hit, Iraq

	Flow Added in Mm^3/yr	Cum. Flow in Mm^3/yr	Percent of Total
Flow at Birecik (1937–1963)		26,990	91.7
Added in Turkey	410	27,400	1.3
Added in Syria			
by Sacir/Sajur	80	27,480	0.4
by Balikh/Culap	190	27,670	0.6
by Khabur	1,780	29,450	6.0
Total added in Syria	2,050		7.0
Total added Syria/Turkey	2,460		8.3
Flow at Hit		29,450	

Source: Kolars (1987).

Note: The purpose of this table is to approximate the various shares of water added to the Euphrates between Birecik and Hit. Clawson (1971) data were used for their length of coverage and seeming internal consistency. In some instances, FAO (1966) data were used for tributaries because they are the only record available. The point made here is to show the relative volumes of water each stream contributes. A discharge value of 29,450 Mm^3 at Hit may be low, but the internal consistency and proportions are more important than the actual value.

be expected in the near future? These questions are not easily answered because of the scattered and widely varying bodies of data and estimates that are available to investigators. The discussion, graphs, and tables that follow will try to give the range of such information, to evaluate available data, and to assess which answers are most likely to be correct.

Relative Shares of Euphrates Water: Birecik, Turkey, to Hit, Iraq

Table 6.1 addresses the amount of water provided by each of the sources mentioned above. Despite the fact that a wide range of

Table 6.2. The Sajur/Sacir River
(yearly average flows)

	Length in km	Flow in m³/s max	min	ave	Flow in Mm³/yr	Data Source
IN SYRIA		25.0	0.5	3.0	94.510	FAO, 24
	48	13.6	0.0	1.9	59.920	SAR, Table 4-1
		—	—	2.8[a]	88.000	USAID, I-184
Average of above				2.56	80.800	
IN TURKEY	60			4.4[a]	138.600	GAP III-27

Source: FAO (1966); SAR (1980); USAID (1980); GAP (1980).

[a]Computed from annual value.

data exists for discharge along the course of the Euphrates, the figures in this table have been chosen for their internal consistency and chronological span. The magnitude of Euphrates flow has been discussed in Chapter 5. At issue here are the relative proportions of that flow. Given a volume of 27,000 Mm³ at Birecik, an additional 410 Mm³ enters the main stream before reaching the Turkish-Syrian border at Karkamis. This is a fairly large amount for such a short distance (approximately 25 km), but it includes the flow of the Nizip and several smaller streams. Precipitation is approximately 400 mm annually; runoff is about 100 mm/yr along this stretch of the river.

The next measured increment of stream flow is from the Sajur, which rises in Turkey and enters the Euphrates a short distance inside Syria on the right bank (Table 6.2). While somewhat greater flow values are shown for Turkey, the diminished downstream flow in Syria can easily be the result of small-scale private irrigation in both countries. A small reservoir reportedly planned for irrigation in Syria might further reduce stream flow through evaporation and extraction (USAID 1980, 1:I–1-VIII-22; RPU 57, page I-184). A small dam and reservoir are planned for the west branch of the Sacir in Turkey (the Tuzel Suyu). The Tuzel has an average annual natural flow of 40.15 Mm³, and the reservoir will have an effective capacity for irrigation purposes during the months of June–July–August of 46.3 Mm³. No indication is available of the area to be irrigated; but, assuming a 50 percent use of the available water (minimum reservoir capacity 5.7 Mm^3), a total of 20.3 Mm^3 could be removed from the Sajur's flow downstream. This is not a significant amount, but it is one which may necessitate international negotiation for the optimum use of this stream by both countries. (All Turkish data from GAP 1980, V-24.)

Continuing downstream, the head of Lake Assad formed by the Tabqa Dam is encountered south of Yusuf Pasha at the village of Remis. This reservoir has a storage capacity when filled to a crest height of 40 m (300 m above sea level) of 11,600 Mm^3 (SCBS 1980) and a surface area of 625 sq km. Loss by evaporation from this surface is significant and will be discussed elsewhere in this study.

An underground aqueduct leads from a pumping station on Lake Assad, southeast of Khafsah Kabir, to the city of Aleppo. This is apparently the major—and perhaps the only—source of water for that city at present. The Qweik River, which rises in Turkey, at one time supplied water for the city of Aleppo. It seems, however, that little or no city water has come from this source since the 1940s (Richard Dekmejian, personal communication, n.d.). Present use amounts to 220,000 m^3/day, which is about 145 liters per capita, for a total of 80.3 Mm^3/yr (USAID 1980, 1:I–1-VIII-22; RPU 20, I-69). It is not the purpose of this section to consider the impact of withdrawals of this nature. However, it is interesting to note that this amount is approximately equal to that added by the Sajur upstream. Domestic demand will soon exceed this amount. At the same time, fruit canning at Idlib, two cement plants, a glass plant, and a sugar factory are all significant water users. Further details of this situation are given in Table 6.3, which describes the Qweik River. While this river does not feed into the Euphrates drainage system, it relates to water use problems common to both systems. Sewage facilities in Aleppo are considered to be "totally inadequate" (RPU 20).

The Balikh (Turkish: Culap) is the next tributary. It enters on the left bank and receives the bulk of its water from the Ain Arous (spring) in Syria, near Tel Abiad on the Turkish border. Additional flow crosses the border from Turkey but the consistency of this is uncertain from the data available (Table 6.4). The 116 km length

Table 6.3. The Qweik/Balik River[a]
(yearly average flows)

Location	Flow in Mm^3/mo			Flow in Mm^3/yr	Data Source
	max	min	ave		
Yagiz Kopru (6 yrs)	5.05	0.30	1.84	22.02	GAP, III-27
Syria	7.0	0.0	0.5	15.8	SAR, Table 4-1
Kemlim Dam[b] site				19.84	GAP, V-24
near Aleppo[c]			2.79	88.0	USAID, I-69

Source: GAP (1980); SAR (1980), USAID (1980).

[a]The Turkish name for this stream is the Balik. This should not be confused with the Syrian name for the Turkish Culap, which is Balikh.
[b]The Kemlim Dam is planned by the Turks for the Balik River. Minimum reservoir capacity 2.78 Mm^3, effective reservoir capacity 31.72 Mm^3. No irrigation hectarage available.
[c]". . . it appears that most of this water is used in Irrigation Network 8 downstream in RPU 26." Network 8 at Matkh has 14,860 ha (USAID 1980).

of the Balikh in Syria (SCBS 1980, Table 4-1) is heavily utilized for irrigation. The same is true for the Culap (Balikh) in Turkey, where the stream and its tributaries are apparently dry for varying periods of time. No data are available in usable form to indicate the exact amount of land irrigated or water used in either country along the Balikh/Culap River system. The quantity must be considerable.

Some estimate of the impact of Turkish use in future years may be made. According to plans, some 160,000 ha will be irrigated on the Urfa-Harran Plain (GAP 1980, V-4). Water for this will come through the Urfa Tunnel from the lake behind the Ataturk Dam. Return flow from these fields may range between 2,300 and 5,800 m^3/ha, depending upon the interpretation chosen for the data (see Tables 7.4 and 10.3). This would increase the flow of the Balikh by amounts ranging from 368 to 928 Mm^3/yr. This would essentially double to quadruple the downstream flow. While this may present new opportunities for irrigation in Syria, the quality of this water may be poor as a result of upstream leaching and/or dissolved fertilizers, herbicides, and pesticides. Flooding might also present problems. Again, while mention

Table 6.4. The Balikh/Culap River
(yearly average flows)

Location	Length of Record	Flow in Mm^3/mo max	min	ave	Flow in Mm^3/yr	Data Source
IN TURKEY						
Incirli	14 yrs	5.09	0.40	2.28	27.39	GAP, III-22
Horozkoy	2 yrs	25.20	0.02	7.96	95.48	GAP, III-22
Kopruluk[a]	2 yrs	4.45	0.09	1.30	15.59	GAP, III-22
SUB-TOTAL[b]					111.07	
IN SYRIA						
Ain Arous	?			15.77	189.22	FAO & USAID
SUB-TOTAL					300.29	
Cermelik Kopru[c]	1 yr	0.25	0.00	0.03	0.38	GAP, III-22
TOTAL[d]					300.65	

Source: GAP (1980); FAO (1966); USAID (1980).

[a]Kopruluk is on the Cavsak tributary in Turkey.
[b]Subtotal is sum of flow of Horozkoy and Kopruluk, two tributaries measured individually. The Incirli measurement is far upstream above Horozkoy.
[c]Cermelik Kopru is on the Karacurum in Turkey, which enters the mainstream in Syria.
[d]Despite this total, the more conservative value based on the flow of Ain Arous (189.22 Mm^3/yr) has been used in Table 6.1 because that is the value reported downstream in Syria.

should be made of these issues, they remain secondary to the main purpose of this section's discussion.

The final contribution to the flow of the Euphrates comes from the Khabur River system, which joins the mainstream 40 km downstream from Deir ez-Zor. The use of this stream in Turkey and Syria and the complexities relating to its various tributaries and ground water resources justify a detailed analysis in the pages that follow. At this point, it is sufficient to say that the "natural" flow of the stream at Suwar is about 56.5 m^3/s (1,780 Mm^3/yr). It should also be noted that, wherever possible, data have been used in the calculations for Table 6.1 that predate major dams and developments along the rivers concerned.

The Relationship Between Euphrates Flow and That of Its Syrian Tributaries

Just as the discharge of the Euphrates varies widely from year to year, the difference in discharge between Birecik, Turkey and Hit, Iraq varies greatly. Sometimes, as in 1941, 1951, and 1959, there was actually less water at Hit than in Turkey. On the other hand, positive increments have varied from as much as 7,600 Mm^3 (in 1954) to as little as 400 Mm^3 (in 1944). The average difference is 2,470 Mm^3 more at Hit than in Turkey, based on 27 years of measurements (1937–1963). These variations are shown by Table 6.5, and by Figures 6.1 through 6.7.

Figure 6.1 shows the incremental differences between discharge at Birecik, Turkey and Hit, Iraq from 1937 through 1963. (This is the longest consecutive record for both gauging stations available for this analysis.) Discharge at Birecik is indicated as a flat line by the abscissa. As can be seen, no particular trend is evident in the variation of these differences. Figure 6.2 shows the correlation—or lack thereof—between the quantity of water discharged at Birecik and the incremental difference recorded at Hit when data from both sites are plotted for the same year.

On the other hand, Figure 6.3 shows a clear positive correlation (r = 0.92) between the total discharge at Birecik and the total discharge at Hit. This indicates that the amount of water discharged across the border from Turkey into Syria will definitely affect the amount of water arriving downstream in Iraq. However, it is the flow of the main stream and not the flow of its tributaries in Syria which underlies this phenomenon. This implies that variations in the flow of the Khabur in Syria, whether from natural or human causes, may increase or decrease the amount of water available in any given year but significant deficits downstream in Iraq are either the result of water removals from the main stream by human action in Turkey or Syria or of major climatic variations in the catchment area in Turkey.

The question of why, in terms of flow, some years are lean and some abundant on the Khabur and/or Balikh remains to be discussed. If these two streams were to be dried up completely,

Table 6.5. Yearly Flows at Birecik and Hit
(in chronological order)

| | At Birecik | | | At Hit | | |
Year	Flow in m^3/s	Flow in Mm^3	2-Yr Ave.	Flow in m^3/s	Flow in Mm^3	Difference in flows
1937	894	28,200		862	27,800	− 400
1938	997	31,400	29,800	1,076	33,900	+2,500
1939	831	26,200	28,800	935	29,500	+3,300
1940	1,165	36,700	31,500	1,182	37,300	+ 600
1941	1,120	35,300	36,000	1,117	35,200	− 100
1942	1,032	32,500	33,900	1,078	34,000	+1,500
1943	856	27,000	29,800	1,023	32,300	+5,300
1944	1,056	33,300	30,200	1,069	33,700	+ 400
1945	691	21,800	27,600	851	26,800	+5,000
1946	920	29,000	25,400	1,047	33,000	+4,000
1947	703	22,200	25,600	809	25,500	+3,300
1948	1,007	31,800	27,000	1,119	35,300	+3,500
1949	662	20,900	26,400	711	22,400	+1,500
1950	753	23,700	22,300	799	25,200	+1,500
1951	716	22,600	23,200	700	22,100	− 500
1952	932	29,400	27,000	963	30,400	+1,000
1953	906	28,600	29,000	1,119	35,300	+6,700
1954	1,012	31,900	30,300	1,254	39,500	+7,600
1955	588	18,500	25,200	710	22,400	+3,900
1956	827	26,100	22,300	876	27,600	+1,500
1957	818	25,800	26,000	893	28,200	+2,400
1958	655	20,600	23,200	744	23,500	+2,900
1959	574	21,300	21,000	638	20,100	−1,200
1960	826	26,000	23,700	973	30,700	+4,700
1961	484	15,300	20,700	535	16,900	+1,600
1962	692	21,800	18,600	749	23,600	+1,800
1963	1,356	42,800	32,300	1,378	43,500	+ 700

At Birecik:	At Hit:	Difference:
N = 27	N=27	x = +2,378
x = 856 m^3/s	x = 934 m^3/s	
x = 27,000 Mm^3/yr	x = 29,500 Mm^3/yr	

Source: Clawson et al. (1971); computations by Kolars.

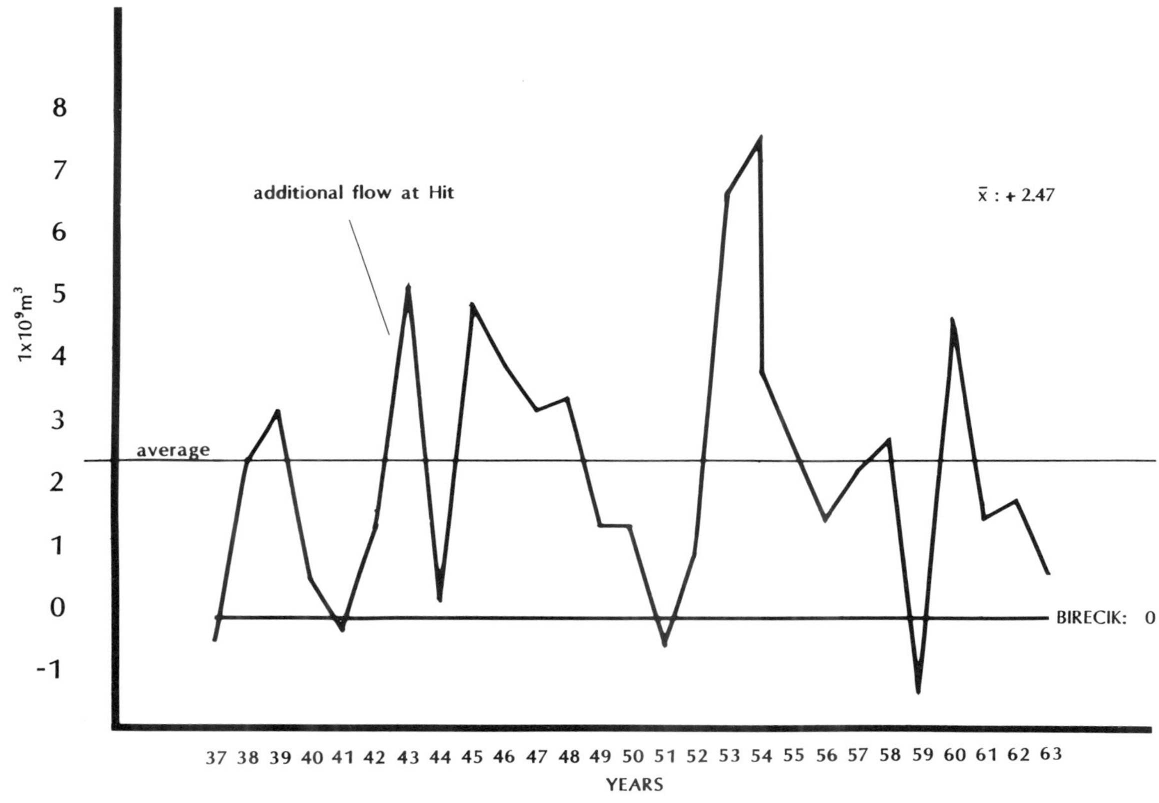

Figure 6.1. Difference in discharge at Birecik, Turkey, and Hit, Iraq: 1937–1963.

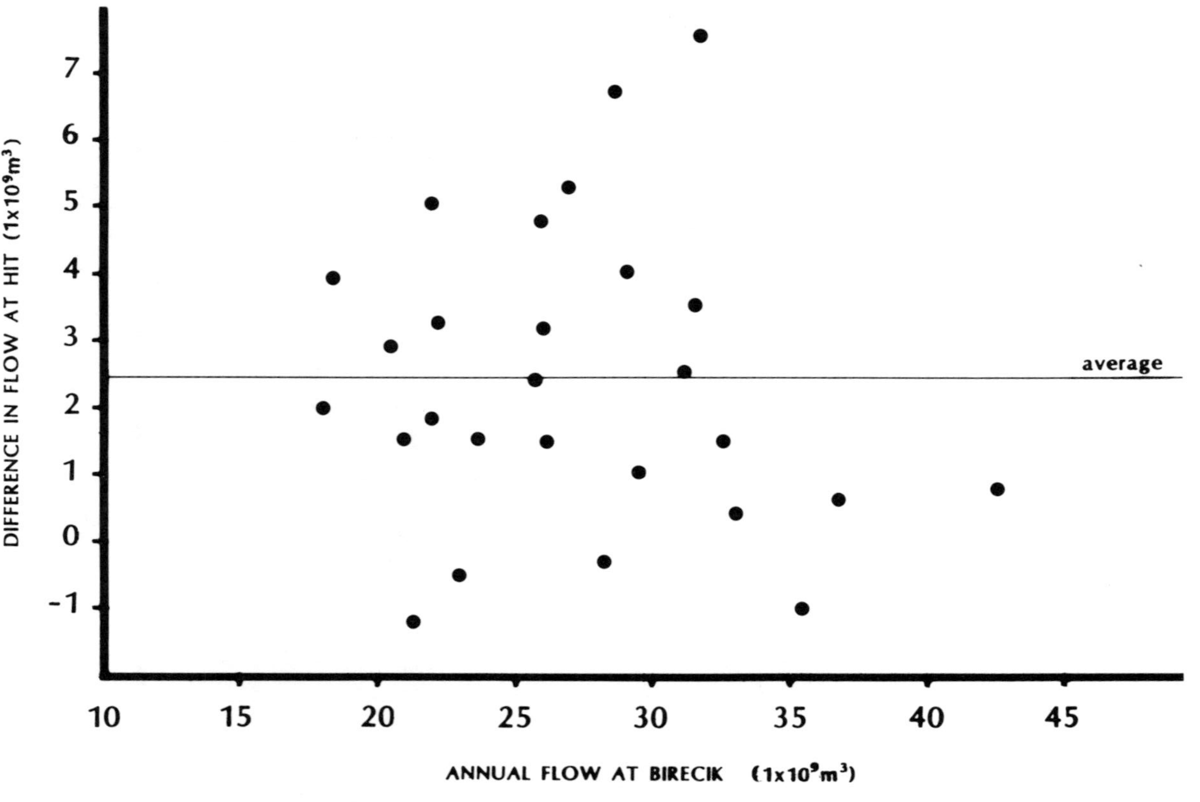

Figure 6.2. Relationship between flow at Birecik and incremental difference in flow at Hit. Source: Clawson; computations by Kolars.

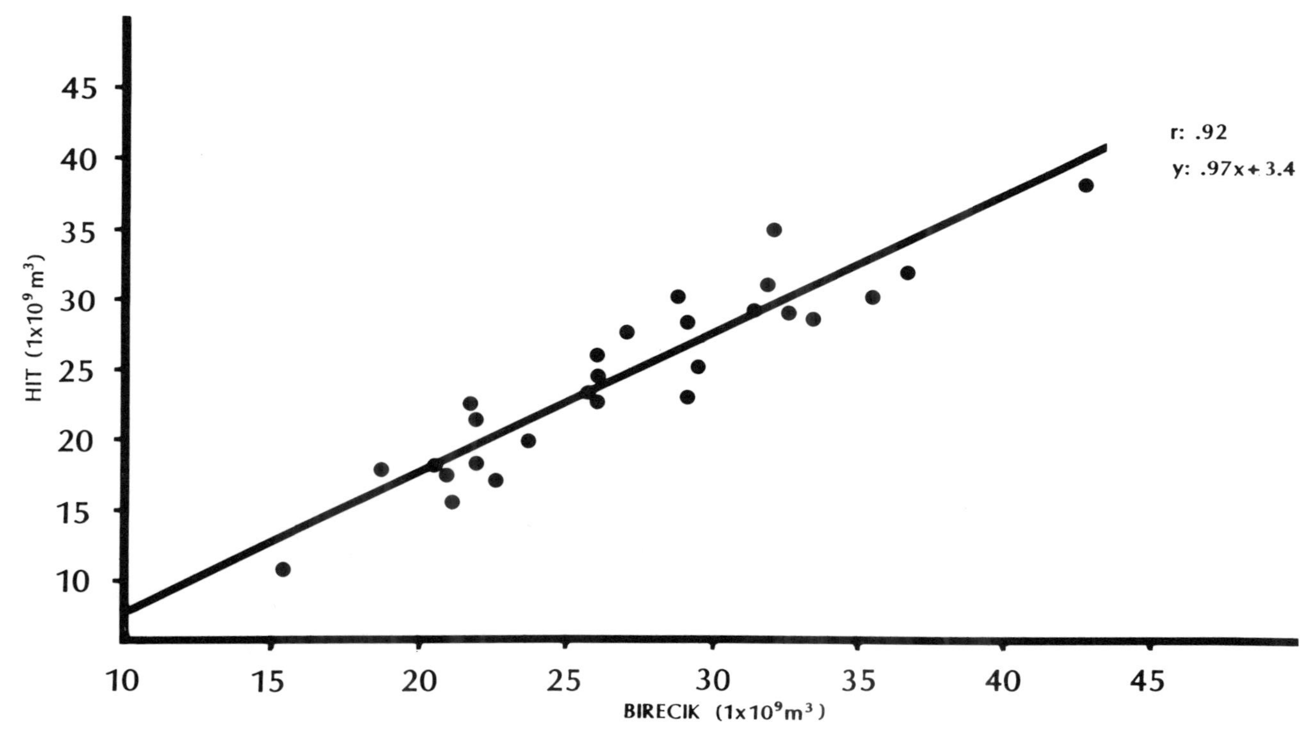

Figure 6.3. Relationship of annual discharges at Birecik, Turkey, and Hit, Iraq, 1937–1963. Source: Clawson.

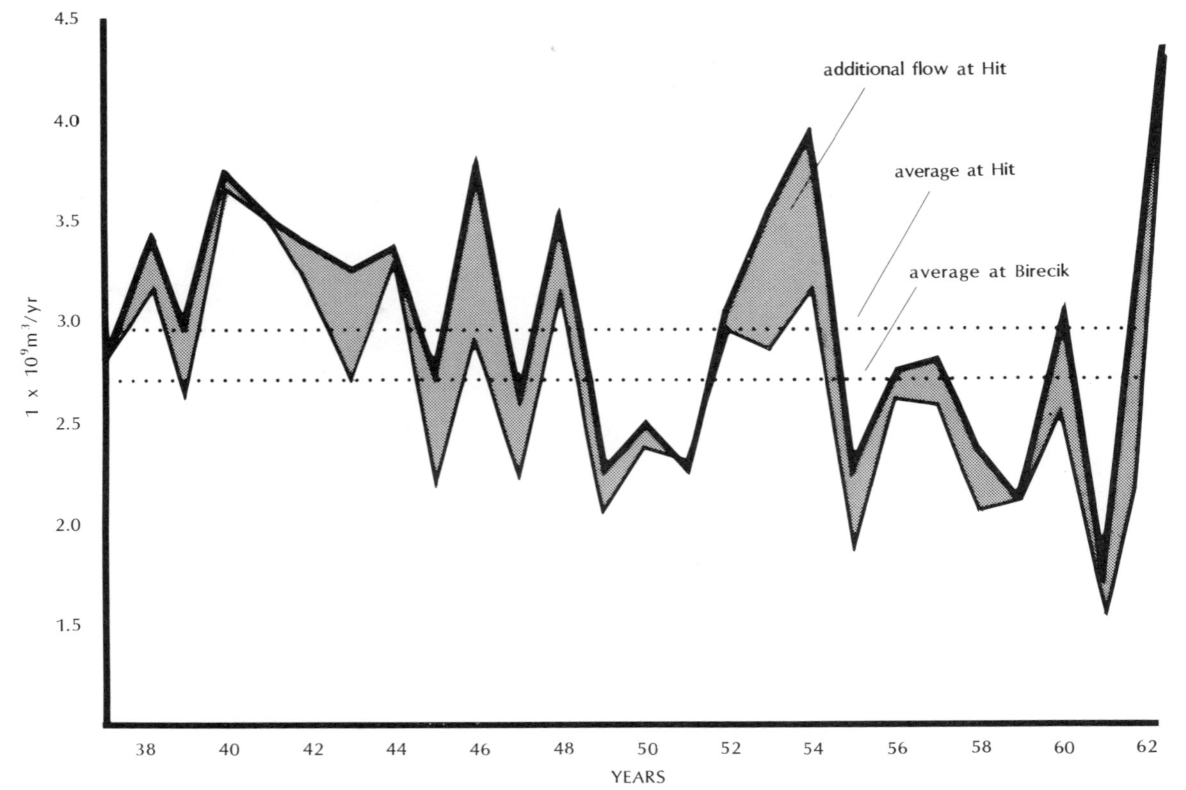

Figure 6.4. Discharge of the Euphrates at Birecik, Turkey, and Hit, Iraq, 1937–1963.

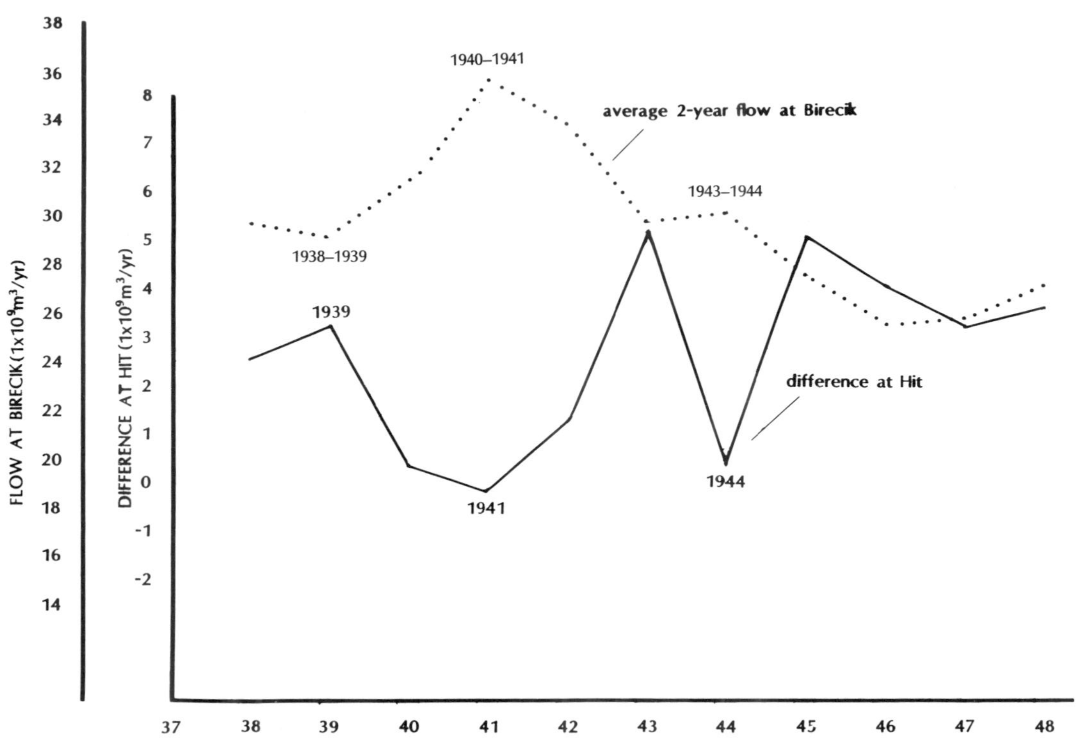

Figure 6.5. Correlation between flow at Birecik and Hit, 1937–1944. Source: Clawson; Computations by Kolars.

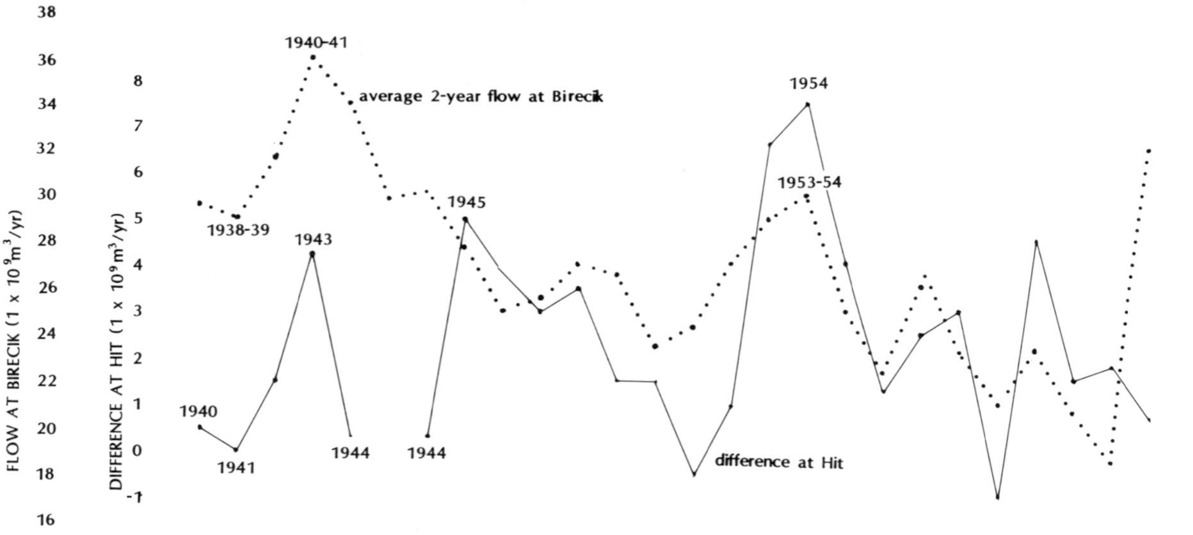

Figure 6.6.   Correlation of 2-year averages at Birecik and Hit, 1937–1963. Source: Clawson; computations by Kolars. N. B. Average increments for 1940–1944 are staggered by two years against Birecik values. Beginning with the second 1944 values coincide annually (e.g. 1953–54/1954).

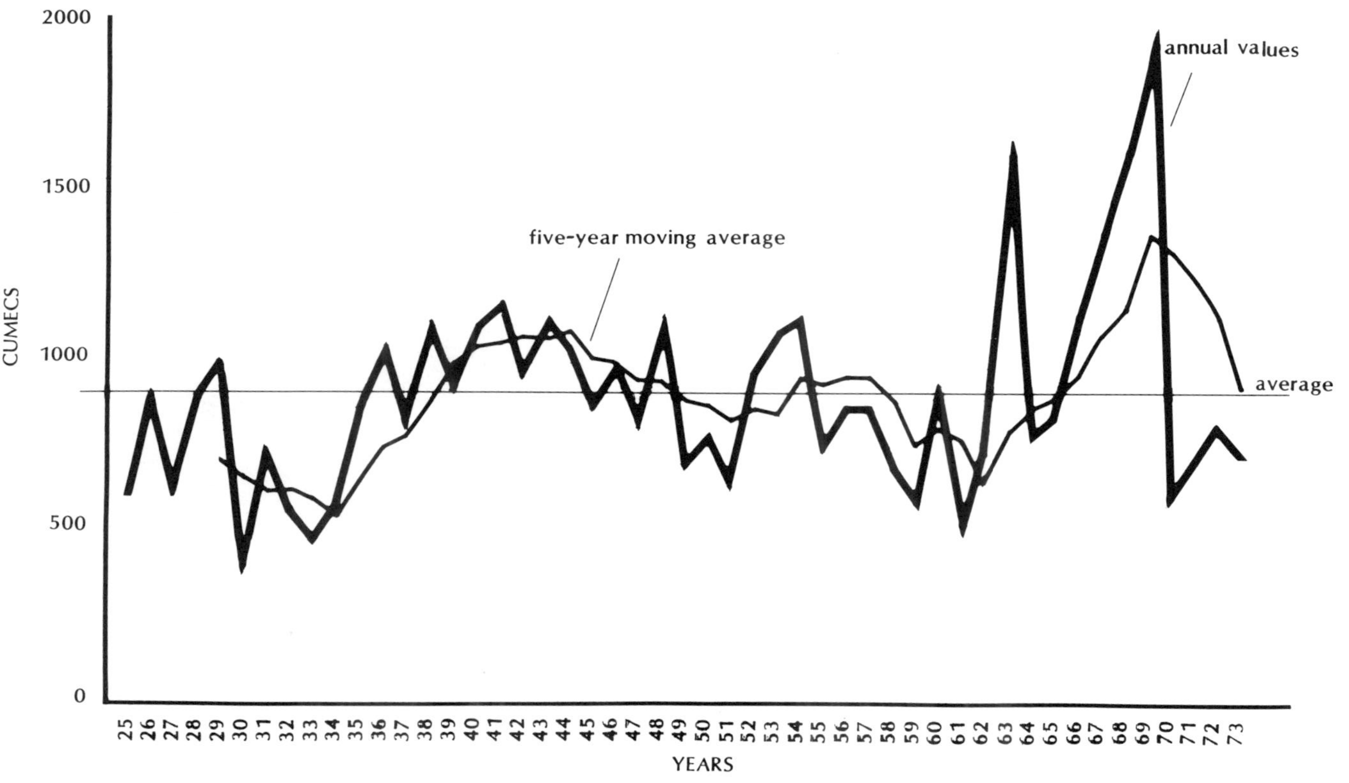

Figure 6.7. Discharge of the Euphrates River at Hit, Iraq, 1924–25/1972–73. Source: Hadithi, Appendix E (Ministry of Irrigation); five-year averages computed for present study.

the flow of the Euphrates would on the average be reduced by 6.6 percent in Iraq, but year-to-year variation in the incremental flow reaching Hit has a more complex relationship to variations in the flow of the Syrian tributaries (Figure 6.4).

The nature of this relationship is shown by Figures 6.5 through 6.7. In the first of these Figures, the discharge of the Euphrates at Birecik has been taken as a general indicator of conditions throughout the region, including the headwaters of the Syrian tributaries. Run-off appears to be a function of the holding capacity and permeability of the soil and perhaps of major underground aquifers. The lack of correlation shown in Figures 6.2 and 6.5 (where mainstream flow peaks appear, if anything, to be diametrically opposed to incremental peaks) can be largely eliminated by taking two-year running averages at Birecik and plotting them against increments at Hit (Figure 6.6). This assumes that one-half the water within the watershed will be retained for a given year and runoff in the next. It should be noted that the correlation is good for the years from 1945 to 1961 when each two-year average is plotted against the same year at Hit, but that for the sequences 1940–1944 and 1962–1963—as shown for the former on Figure 6.5—the diametric opposition of flows to increments is accentuated by the averaging process. This, in turn, has been overcome (Figure 6.6) by staggering the downstream values by two years (for example, 1941 is correlated with 1938–1939). Before suggesting an explanation of this delayed arrival downstream, it is reasonable to suggest that, given the flow at Birecik in Turkey, it should be possible to predict the "natural" flow of the tributaries in Syria. This may be of considerable importance in the future. (The one caveat to this statement is that the discharge data at Birecik must be accurately equated with "natural" flow.)

Figure 6.7 suggests an explanation of the incremental lag described above. This shows the measured flow at Hit for each year, beginning in October and ending with September of the following year, for the period 1924–1925 through 1972–1973. (A similar time span for Birecik is, unfortunately, unavailable, so some of the ideas that follow must remain as untested speculations.) Inspection shows that the lag period 1940–1944 (Figure 6.6) followed the severe drought of 1930–1936 (Figure 6.7). A critical transition year was 1945, when the lagged arrival of the increment ended and a year-to-year correspondence began. This was the eighth year after the drought that flow had been above average (as shown by the five-year running average also plotted on Figure 6.7). This suggests that considerable time is necessary to recharge groundwater reservoirs before they are full enough to allow an added increment to be passed downstream in the same year.

The impact of excess runoff is less certain, and, because of the shortness of the available record, it is not possible to test the effect of the great discharge of 1969 against later years. Some lag effect is indicated for the years 1957–1958 (Figure 6.6) following the heavy discharge at Birecik in 1954, but the data do not warrant much speculation. Nevertheless, the above discussion allows a clearer view of the situation on the tributaries in Syria. As mentioned earlier, the complexity of the Khabur system and the emphasis placed on its future development by the Syrians justifies a detailed examination in subsequent chapters.

7

Water Use per Hectare and Anticipated River Depletion

It is of particular importance to establish a reasonable expectation of water use per hectare of farmland (irrigated) in the GAP area. By extension, this discussion can be applied to similar circumstances in Syria. (Iraq with its severe drainage problems leading to salination requires separate consideration.) As has been alluded to in Chapter 4, the amount of water used on 1 square mile (or on each hectare) of irrigated cropland can be significant. It becomes important, therefore, to ascertain just how much will be used by Turkish and Syrian irrigation developments. If a crop requires 1 m of water per growing season instead of only 500 mm, water demands will be doubled.

Moreover, the very large hectarages anticipated by projects in both countries mean that immense quantities of water are at stake. Several agencies have published computations showing the anticipated water needs for various locations during the warm months when plant growth is possible. Because several methods exist for computing such needs, because published reports often give conflicting values for the same or similar locations, and because in the Turkish case the definition of "irrigation water used" is unspecified, it becomes necessary to consider carefully the meaning and accuracy of such data.

Because temperature increases rapidly from north to south in the Euphrates river basin and because precipitation decreases in the same direction the demands of irrigation vary from site to site, as does the supply naturally available for plant growth. The critical measure in this case is *potential evapotranspiration* (PE). This refers to the theoretical amount of water a field crop needs in its immediate surroundings to meet evaporation demands from its own surfaces and from the ground upon which it grows and also to meet its own transpiration demands in order to ensure a healthy metabolism throughout its growing season.

Potential evapotranspiration is calculated in several ways, each beginning with monthly and annual air temperatures. More complex methods include wind speed and other criteria. Evapotranspiration measures are usually computed for an entire year, although the growing season (in the case of the Euphrates roughly from April through October) is the period which interests us here.

Methods of Computation

Two methods have been used to compute water needs by the various authors of the source materials referred to in this report. The two are the Blaney-Criddle formula (GAP 1980, III-20) and the Thornthwaite method.[1] Both use day-length and temperature as major independent variables. Thornthwaite's method does not refer to crop type, while Blaney-Criddle's does by reference to an empirical crop factor "k" which varies with crop type and stage of growth. Neither method takes existing precipitation or soil moisture into account. Part of a crop's (or plant's) PE need will be met with water supplied by natural precipitation and/or water stored in the soil. The amount thus supplied without supplemental irrigation is referred to as *actual evapotranspiration*, which in

---

1. For a description of these two methods and a comparison of them with the Penman method, see Dunne and Leopold (1978, 136–141). Computations by the author of this text were based on Thornthwaite's Water Balance for two reasons: the data (air temperature and precipitation) were available where other measures (e.g., wind velocity) were not and the Water Balance takes precipitation and groundwater into account, thus presenting a more realistic view of the agricultural process. Calculations were based on Thornthwaite and Mather (1957, vol. 10, no. 3).

Table 7.1. Irrigation Water Needs—"Sulama Suyu Gereksinimi"
(Mardin-Ceylanpinar)

	m^3/ha/mo
April	405.34
May	832.87
June	2,090.56
July	2,890.21
August	2,438.08
September	1,169.28
October	172.37
TOTAL	9,998.71

Source: GAP (1980, III-36).

tal irrigation is referred to as *actual evapotranspiration*, which in arid regions may be significantly less than the PE. What remains is the "water deficit" (D), which must be compensated for by irrigation.

Thornthwaite (1957) subsequently devised a method of computing the "Water Balance" for a given crop area. With this method, available moisture—either as precipitation or as soil moisture—is subtracted from the potential evapotranspiration need computed for a given area with a particular soil type (sand, loam, clay), temperature, and crop (deep rooted, shallow rooted, etc.).[2] In this way, the amount of water needed to be supplied by irrigation can be computed.

Table 7.1 illustrates the type of data available from Turkish sources for some, but not all locations. Given a reasonable distribution of such data sites, extrapolations between them for the entire area are possible. Another source of water use data consists of values calculated using one of the methods described above.

2. It should be noted that Thornthwaite's method tends to underestimate need while the Blaney-Criddle method is somewhat more exact. The Thornthwaite method was used herein out of necessity (see note above). On the other hand, such low estimates may be taken to represent the absolute minimum amount of water necessary, thus establishing a baseline for discussion purposes.

Turkish sources present computations based on the Blaney-Criddle method (GAP 1980, III-20). The Thornthwaite method has been used here to check such values.

The first question to be asked is what is meant by "irrigation water needs," the direct translation of the Turkish phrase quoted in Table 7.1. Both Blaney-Criddle and Thornthwaite equate their formulas with the *potential evapotranspiration needs of the crop*. This refers directly to the amount of water a field crop needs but does not take into account precipitation and soil moisture which may be available. The Turkish usage might mean one of three things:

1. potential evapotranspiration only
2. total amount withdrawn from the reservoir—which would include potential evapotranspiration, water losses resulting from system inefficiency, and the amount of water which eventually finds its way back into the system farther downstream
3. potential evapotranspiration plus the amount lost to system inefficiency but excluding the amount returned to the system

These three possibilities are shown in Table 7.2.

Definition of Components

As Dunne and Leopold (1978, 162) point out, significant additional water loss beyond evapotranspiration needs occurs during transfers from reservoir to the farm and from the main canal to individual fields. They suggest that, as a rule of thumb, evapotranspiration needs should be doubled to account for such losses. This problem, in terms of Turkey and Syria, is discussed elsewhere in this study, but for this analysis it is taken to be 2.5 times the evapotranspiration. At this juncture, 35 percent of the total amount withdrawn from the reservoir is assumed to be "return flow" to the Euphrates at some point in the system. The components considered by Table 7.2 are:

Table 7.2. Interpretations of "Suluma Suyu Gereksinimi"
(irrigation water needs)[a]

[b]Stated Value 9998.71 (10,000)	Interpretation/ Explanation	Amount Withdrawn [W] 2.5 PE	Amount Returned [R] 0.35 (2.5PE)	Potential Evapotransp [PE] (stated/comp)	Water Loss [L] W−(PE+R)	Fund Depletion [FD] L+PE
1. 10,000	Potential Evapotranspiration Only PE	25,000	8,750	10,000[c] (9998.71)	6,250	16,250
2. 10,000	Total Amount Initially Withdrawn PE+L+R=W	10,000 (9998.71)	3,500	4,000	2,500	6,500
3. 10,000	Potential Evapotranspiration Plus Amt Lost (Excludes Amt Returned) PE+L=FD	15,385	5,385	6,154	3,846	10,000 (9998.71)

Source: GAP (1980).

[a]Time period = April through October. All values in m^3/ha.
[b]As stated in GAP, computed by Blaney-Criddle method; "k" unspecified in text.
[c]Cf. calculated PE (Apr-Oct) in Table 7.3 (Ceylanpinar).

- *stated value*—a value given without definition in the Turkish example (for example, just what is meant by the term "irrigation water need" and by the figure 9998.71 m^3/ha/April-October?)
- *amount withdrawn*—the quantity of water which would actually be withdrawn from the reservoir given a particular definition of the first term—2.5 times deficit replacement (once deficit replacement has been determined. See definition below.)
- *amount returned*—amount that eventually ends up back into the river system. It is assumed that 35 percent of all water drawn from the reservoir will eventually find its way back into the river system. This is often referred to as *return flow*.[3] (See Appendix A for how this value was determined.)
- *potential evapotranspiration*—the amount of water required as defined in the preceding text during the growing season April through October.
- *water deficit or deficit replacement*—that portion of the potential evapotranspiration which cannot be made up by precipitation or soil moisture and must be met by added irrigation water. (This term is used in Table 7.4.)
- *water loss*—that portion of the water withdrawn from the reservoir which neither returns to the river (return flow) nor is used to satisfy *deficit replacement*. This disappears through seepage, evaporation from canal surfaces, evapotranspiration from wild vegetation, etc.
- *fund depletion*—the amount withdrawn from the reservoir less return flow. In other words, the absolute drain on the river system (measured per hectare of irrigated land) which diminishes downstream flow. This would consist of "water loss," as described above, plus the "deficit re-

---

3. This study recognizes that water lost through system inefficiency may be returned to its original source (i.e., river). However, in the cases under discussion water is being delivered and applied during months where there is a significant water deficit and no precipitation. Moreover, field conditions are *not* confined to well-defined main river drainage systems which might readily regain lost water through underground flow. Therefore, return flow (RF) and losses due to system inefficiencies are treated separately herein.

placement" which is used to supply direct crop/plant needs unmet by precipitation or groundwater.

Potential Evapotranspiration

Row 1 of Table 7.2 assumes that the figure quoted in the Turkish source (9998.71—here rounded off to 10,000 for convenience) represents potential evapotranspiration (PE) for the period April through October. The total amount withdrawn given the criteria described above would be 25,000 m³/ha, of which 16,250 m³/ha would constitute an absolute loss to the system (diminishing downstream flow for use in Syria and/or Iraq).

Row 2 assumes that the 10,000 m³ quoted refers to the total amount withdrawn for all purposes. This would allow potential evapotranspiration of only 4,000 m³/ha during the entire growing season and can be dismissed as unrealistically low.

Row 3 assumes that the 10,000 m³ refers to the fund depletion (that is, the amount lost absolutely to downstream flow). This would allow 6,154 m³/ha for April through October. While this might be a possibility, the PE was recalculated using the Thornthwaite method and Turkish temperature and precipitation data. The result is 9730 m³/ha for April/October, as shown in row 6, Table 7.3. Since the Thornthwaite method results in *lower* estimates than does the Blaney-Criddle method, which the Turks used, it is obvious that row three does not offer the correct definition of the term in question.

The above discussion constitutes a tortuous, but necessary checking of the meanings used. *It may be assumed that the Turks are referring to potential evapotranspiration alone for the months April through October.* It now becomes possible to assign evapotranspiration values elsewhere in the river basin and to consider the water deficit or deficit replacement in terms of the water balance, a more realistic measure of the basic water needs of the various irrigation projects planned for Turkey and Syria.

Table 7.4 lists the potential evapotranspiration rates published for various locations in the two countries. Attention should be given to the top row, which lists annual temperatures from south to north. This gives a good indication of the relative standing of the various stations involved. Because PE is a function of temperature and day length, it is logical to expect diminishing water needs as annual temperatures decline. (See Figure 7.1 for a diagrammatic view of these relationships.)

Values given in the United Nations Food and Agriculture Organization survey (FAO 1966) of the Khabur region are consistent with our expectations. Penman values in row 3 are higher than those derived from the Thornthwaite method, a fact again consistent with the two techniques. An anomaly exists with the GAP (1980) data. The PE cited for Ceylanpinar is greater than that given for Nusaybin, although the annual temperature for the latter is higher. On the other hand, values calculated for this study using the Thornthwaite method show a consistent diminishing from south to north (Tables 7.5–7.8). Thus, FAO data and those derived for the present study are preferred to the ones given in GAP.

Water Balance

A more meaningful value for water use is shown at the bottom of Table 7.3. The water balances as given by the FAO (1966) for Syrian stations and as computed for Turkish stations in this study show a consistent decline from south to north. Moreover, these values take into account the precipitation and groundwater available during the entire growing season for each station. (A soil moisture retention of 200 mm was assumed for the Turkish calculations.) It should be noted that the reversal of values for Nusaybin and Ceylanpinar in these data is consistent with the greater rainfall at the former location. (This may account for the inconsistent reversal of the Turkish data mentioned above, if the "k" values used in the Blaney-Criddle method took this into account through either plant type or time within the season; but, since there is no explanation of the technique used, the GAP (1980) data must still be treated with caution.)

It is important to note at this point that values for the Thornthwaite (1957) water balance are only 70 percent of the values cited in GAP for the same stations. Despite the fact that Thornthwaite

Table 7.3. Potential Evapotranspiration: Turkish and Syrian Locations

Locations	Deir ez-Zor	Tel Tamir	Qamishli/ Nusaybin	Ras al-Ayn/ Ceylanpinar	Urfa	Siverek
Annual Precip[a]	(148 mm)	(300 mm)	(452Q/463N mm)	(R-A 315 mm est.)	(462 mm)	—
Annual Temp (C) GAP	—	—	18.9	18.2	18.0	16.2
Annual Temp (C) Map #1 (pocket), FAO	20.00*	18.0**	19.3*	"<18"**	18.1**	—
PE[f] Apr-Oct Penman Method FAO, 62	1,302	—	1,193	—	—	—
PE[f] Apr-Oct Thornthwaite Method FAO, 61	1,128	—	1,121	—	—	—
PE[g] Apr-Oct "Sulama Suyu Gereksinimi" GAP, III-36	—	—	9,805[b]	9998.7[b]	8920.1	10461.3[b]
PE[g] using Thornthwaite Method GAP T&P data Apr-Oct	—	—	9,984	9730[c]	9649	8811

Water Balance[g] Deficit Using Thornthwaite Method FAO, 62	10,360	7720	—	—	—	—
Water Balance[g] Deficit Using Thornthwaite Method[e]	—	—	6910[d]	7070[d]	6618	5510

Source: FAO (1966); GAP (1980).

[a]Precipitation as per FAO Map I (pocket).
[b]As stated, but questionable (i.e., out of sequence with M-S temperature sequence).
[c]This figure, being lower than that given by GAP, is consistent with the difference between Penman's and Thornthwaite's methods.
[d]The reversal of the logical sequence (based on temperature alone) results from greater annual precipitation at Qamishli-Nusaybin (485 mm) than Ras al-Ayn-Ceylanpinar (328 mm).
[e]Based on soil moisture retention of 200 mm.
[f]Values in mm.
[g]m^3/ha/growing season.
*1950–1960.
**1957–1960.

Table 7.4. Annual Water Fund Depletion
(m^3/ha/yr and total irrigated area/Mm^3/yr)

Location	Deficit Replacement m^3/ha [D] See Table 7.3	Amount Withdrawn $2.5 \times D = W$	Amount Returned $0.35 \times W = R$	Water Loss $W-(D+R)=L$	Fund Depletion m^3/ha $D+L=FD$	Area To Be Irrigated ha	Total Fund Depletion Mm^3	Total Returned To System Mm^3
Siverek GAP V-4	5,510	13,775	4,821	3,444	8,954	147,000	1,316.2	708.7
Urfa GAP V-4	6,618	16,545	5,791	4,136	10,754	136,000	1,462.5	787.6
Birecik ??? 1984	est. 6,500	16,250	5,688	4,062	10,562	92,700	979.1	527.2
Nusaybin GAP V-4/ Qamishli	6,910	17,275	6,046	4,319	11,229	47,000	527.8	284.2
Ceylanpinar GAP V-4/ Ras al-Ayn	7,070	17,675	6,186	4,419	11,489	UPPER 206,000 LOWER 164,000 FROM AQUIFER (60,000)	2,366.7 1,884.3 (689.3)	1,404.1 1,117.8 (371.2)

				TOTAL FROM CANALS	792,700	8,536.5	4,829.6
				TOTAL FROM AQUIFER	(60,000)	(689.3)	(371.2)
				TOTAL	852,700	9,225.8	5,200.8

SYRIAN VALUES

| Tel Tamir | 7,720 | 19,300 | 6,755 | 4,825 | 12,545 | (For Syrian totals |
| Deir ez-Zor | 10,360 | 25,900 | 9,065 | 6,475 | 16,835 | see next section) |

Source: GAP (1980).

Note: The list of projects and locations given here is incomplete. For a total accounting see Table 10.4 Figures are based on deficit computed according to the Thornthwaite method (see Table 7.2).

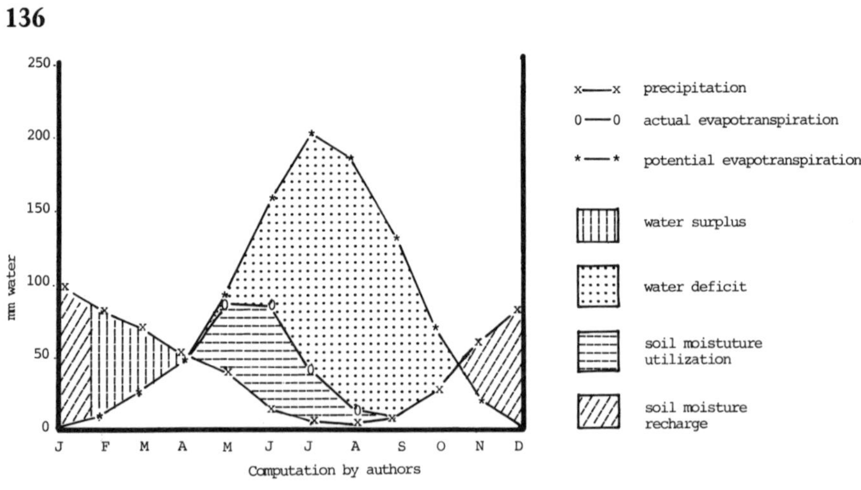

Figure 7.1. Water balance for Siverek (37° 50′ N. Lat.).

underestimates PE compared to the Blaney-Criddle method, the difference—even if only partially accepted—still represents a significant saving in water, if the farm/irrigation managers *carefully follow the water balance method of applying water to their fields and do not over irrigate,* a common failing in such situations.

Given the amounts of water necessary to make up the seasonal deficit, there remains the question of how much water each hectare will require when deficit and water loss are both considered. Also, the question of absolute hectarage planned leads to estimations of total loss to the system.

Table 7.4 provides information regarding total water demand from irrigation in Turkey and Syria. Beginning with Siverek in the north, five locations in Turkey and four locations in Syria allow a transect of the major areas where irrigation is planned. (Two locations in Syria and Turkey, Nusaybin/Qamishli and Ceylanpinar/Ras al-Ayn, share single values.) Total water demand (fund depletion per hectare times total hectarage) is omitted from this table for Syria and will be considered in the section that follows. Total water demand for Turkey is given in an abbreviated form and is discussed more completely in other sections of this study.

Table 7.5. Water Balance for Siverek (37°50′ N)
(Thornthwaite method)

Month	Jan	Feb	Mar	Apr	May	Jun	Jul	Aug	Sep	Oct	Nov	Dec	Yr
T°(C)	2.8	4.4	8.9	13.8	19.7	25.8	30.2	29.6	24.8	18.1	11.0	5.4	16.2
I	.42	.82	2.39	4.65	7.97	11.99	15.22	14.77	11.30	7.01	3.30	1.12	82.2
Unadj. PE	.1	.2	.7	1.4	2.7	4.2	5.4	5.3	4.0	2.3	1.0	.3	
PE	2.6	5.0	21.6	46.2	99.6	156.2	202.5	186.0	124.8	65.8	25.2	7.5	943.0
P(mm)	92.9	81.1	76.9	65.4	44.2	7.5	1.5	1.0	3.0	31.5	60.4	82.5	547.9
P–PE	90.3	76.1	55.3	19.2	−55.4	−148.7	−201	−185	−121.8	−34.3	35.2	75.0	−395
AP WL				0	−55.4	−204	−405	−590	−712	−746			
ST	200	200	200	200	151	71	26	10	5	5	40	115	
Δ ST	85	0	0	0	−49	−80	−45	−16	−5	0	35	75	
AE	2.6	5.0	21.6	46.2	93.2	37.5	46.5	17	8	31.5	25.2	7.5	392
D	0	0	0	0	−6.4	68.7	156.0	169.0	116.8	34.3	0	0	551
S	5	76	55	19									155
RO	2	39	47	33	17	7	4	2	1				155

Soil moisture = 200 mm.

See also Figure 8.1.

Table 7.6. Water Balance for Urfa (approx. 37°10′ N)
(46-yr period, Thornthwaite method)

Month	Jan	Feb	Mar	Apr	May	Jun	Jul	Aug	Sep	Oct	Nov	Dec	Yr
T° C	5.0	6.5	10.2	15.7	21.7	27.7	31.6	31.2	26.6	19.9	13.1	7.3	18.0
I	1.00	1.49	2.94	5.65	9.23	13.36	16.30	15.99	12.56	8.10	4.30	1.77	92.69
Unadj. PE	.1	.3	.7	1.6	3.0	4.8	5.7	5.6	4.5	2.6	1.1	.3	
PE	2.6	7.7	21.6	52.8	109.8	177.1	213.8	196.6	139.1	75.7	27.6	7.5	1031.9
P(mm)	99.8	69.7	64.2	55.4	26.3	2.6	0.5	0.6	1.2	22.1	42.4	85.3	470.1
P–PE	97.2	62.0	42.6	2.6	−83.5	−174.5	−213.3	−196.0	−137.9	−53.6	14.8	77.8	−561.8
AP WL				0	−83.5	−258.0	−471.3	−667.3	−805.2	−858.8			
ST	192.8	200.0	200	200	131	54	18	7	4	3	17.8	95.6	
Δ ST	97.2	7.2	0	0	−69	−77	−36	−11	−3	−1	14.8	77.8	
AE	2.6	7.7	21.6	52.8	95.3	79.6	36.5	11.6	4.2	23.1	27.6	7.5	370.1
D	0	0	0	0	−14.5	−97.5	−177.3	−185.0	−134.9	−52.6	0	0	−661.8
S		54.8	42.6	2.6									100
RO		27	35	19	10	5	2	1	1				100

Soil moisture = 200 mm.

PE of Apr–Oct = 964.86 = 9649 m^3/ha.

Table 7.7. Water Balance for Ceylanpinar (37° N)
(Thornthwaite method)

Month	Jan	Feb	Mar	Apr	May	Jun	Jul	Aug	Sep	Oct	Nov	Dec	Yr
T° C	5.4	7.2	11.1	16.0	22.5	28.7	32.1	31.0	25.6	19.1	11.9	7.2	18.2
I	1.12	1.74	3.34	5.82	9.75	14.09	16.70	15.84	11.35	7.61	3.72	1.74	93.32
Unadj. PE	.2	.6	.8	1.7	3.2	5.1	5.8	5.6	4.2	2.3	.9	.3	
PE	5	15	25	56	117	188	218	197	130	67	23	7	1,048
P(mm)	63	49	46	44	23	1	trace	0	1	16	26	59	328
P–PE	58	34	21	−12	−94	−187	−218	−197	−129	−51	3	52	−720
AP WL			(−32)	−44	−138	−325	−543	−740	−869	−920			
ST	115	149	170	160	99	39	13	5	3	2	5	57	
Δ ST	58	34	21	−10	−61	−60	−26	−21	−2	−1	3	52	
AE	5	15	25	54	84	61	26	21	3	17	23	7	341
D	0	0	0	−2	−33	−127	−192	−176	−127	−50	0	0	−707
S													
RO													

Soil moisture = 200 mm.

Soil moisture cap = 200 for silt-loam (average) for corn, cotton, tobacco, cereals.

Table 7.8. Water Balance for Nusaybin (37° N)
(Thornthwaite method)

Month	Jan	Feb	Mar	Apr	May	Jun	Jul	Aug	Sep	Oct	Nov	Dec	Yr
T° C	5.8	7.6	11.6	16.4	22.9	28.8	32.5	31.5	27.3	20.9	13.6	7.8	18.9
I	1.25	1.89	3.58	6.04	10.01	14.17	17.01	16.23	13.07	8.72	4.55	1.96	98.48
Unadj. PE	.2	.3	.7	1.7	3.4	5.1	5.8	5.7	4.7	2.3	1.1	.3	
PE	5.16	7.65	21.6	56.1	124.4	188.2	217.5	200.1	145.2	66.9	28.1	7.47	1,069
P(mm)	93.9	75.3	68.8	72.2	37.5	1.2	.7	0	.5	15.8	38.2	80.9	485
P-PE	88.7	67.6	47.2	16.1	−86.9	−187	−217	−200	−145	−51.1	10.1	73.4	−584
AP WL				(0)	.87	−274	−491	−691	−836	−887			
ST	174	200	200	200	128	50	17	6	3	2	12	85	
Δ ST	89	26	0	0	−72	−78	−33	−9	−3	−1	10	73	
AE	5.2	7.7	21.6	56.1	110	79.2	33.7	9.0	3.5	16.8	28.1	7.5	378
D	0	0	0	0	15	109	184	191	142	50	0	0	691
S		42	47	16									105
RO		21	34	25	13	6	3	2	1	1			105

Soil moisture = 200 mm.

PE Apr–Oct.

Computations of the water balance for four Turkish stations are shown in Tables 7.5 through 7.8.

Column 2 lists the water deficit per hectare for each location. (Note that the value for Birecik is an extrapolated value.) As discussed elsewhere the amount withdraw from reservoirs will be 2.5 times the stated deficit per hectare (column 3).[4] The amount of the water which reenters the river system is assumed to be 35 percent of that withdrawn (column 4). The water loss per hectare is the total amount withdrawn less the amount returned and the deficit replacement (column 5). The total amount of water per hectare disappearing from the system not to be returned is the fund depletion shown in column 6.

Each of these values can be multiplied by the hectarage found near the station listed in column 1. The results are given for the total fund depletion and for the total returned to the system. *Because these values are based on Thornthwaite's method, which underestimates PE compared with Penman's or Blaney-Criddle's methods, these figures should be considered as minimal conservative estimates of fund depletion and return flow, all else being equal.* The value for 60,000 ha near Ceylanpinar which will be irrigated by water pumped from the aquifer supplying the Ras al-Ayn (springs) is shown separately in parentheses. However, this water, which contributes to the flow of the Khabur in Syria will still have its impact downstream either through reduced flows (total fund depletion) and/or water quality (return flow).

Even this partial listing of projects indicates that, if the Turks were to irrigate 792,700 (+ 60,000) ha from the Euphrates River, this would result in an absolute depletion of 8,500 Mm^3 (+ 700 Mm^3) and a return flow essentially down the Balikh and Khabur

---

4. An independent check on these figures is provided by data relating to Soviet irrigation practiced in Uzbekistan, a temperate desert area. Micklin (1986) reports that "the implied withdrawal rate in 1978 was 15,436 [m^3/ha]." Micklin refers to Lapkin et al. (1981) and *Statistika* (1979, 240). Column 3 shows withdrawals ranging from 13,625 at Siverek to 17,635 at Ceylanpinar/Ras al-Ayn and 25,900 m^3/ha at Deir ez-Zor. Considering the more northerly latitude of Uzbekistan and its shorter summers, the value cited by the Soviets falls within this range.

systems of 5,200 Mm^3. This, in addition to evaporation from reservoirs and additional water use from smaller projects, would have a significant impact upon the downstream river system. An accounting of water uses based on the complete inventory of projects is found in Chapter 10, with further refinement given in Chapter 11.

8

Irrigated Agriculture in the Syrian
Euphrates Drainage Basin

As Turkey withdraws more and more water from the Euphrates River, there will be an increasing deficit downstream, first in Syria and then in Iraq. The ambitious plans of the Syrians to develop their portion of the Euphrates Valley through irrigation will further remove vital water from the stream. The urgency felt by all parties regarding this matter can be illustrated by a seeming contradiction apparent in negotiations among the three riparian users.

As discussed elsewhere (Chapter 2 and Naff and Matson 1984, 90), Iraq apparently has requested a guaranteed flow of 500 m^3/s downstream across its border from Syria. More recently, Syria and Turkey have negotiated a similar flow of 500 m^3/s downstream across the Turkish border into Syria (*Cumhuriyet*, 18 July, 1987). Granted that the side streams entering the Euphrates in Syria on the average increase river flow by approximately 6 percent (the Khabur contributes an average 1,600 Mm^3/yr), a significant discrepancy between the two values must inevitably occur as a result of Syrian needs. It is for this reason that the analyses found in this chapter and Chapter 9 have been made. Before any realistic and binding agreements can be reached among the three users, it is necessary to know what Syria can reasonably be expected to extract from the Euphrates. This, in turn, depends upon the area anticipated for irrigation development as well as the evapotranspiration demands already discussed.

An examination of this topic presents serious difficulties, not the least of which is dated, contradictory, and scarce information. There are four categories of investigation: where and how much land was originally proposed for irrigation, where and how much land did subsequent revisions deem irrigable, where and how much land has actually been prepared for irrigation through state-run projects, and where and how much irrigated land have private farmers and entrepreneurs brought under cultivation?

The first is a matter of record and can be spoken of with some confidence. The second presents a less clear picture but can be estimated with a certain amount of research. The third becomes much more a matter of hearsay dependent upon contradictory sources of information. Moreover the amounts are so small that, though apparently correct, they are given with some hesitation. The fourth involves problems inherent in the data available for examining private activities which make their disaggregation difficult. In the last two instances the data are from four to fourteen years old. Despite such caveats the picture which emerges does allow projections of water use to be made for the long term.

Background to the Problem

Prior to 1950 the waters of the Euphrates were little used. Traditional lifts, often camel powered, brought what little water reached fields on the river's banks. Following independence, however, speculation in cotton by Syrian merchants led to a rapid increase in the number of gasoline pumps drawing water from the river. The amount of irrigated land along the Orontes, the Euphrates, and its tributary the Khabur increased from 284,000 ha in 1956 to 583,000 ha in 1957 (Sanlaville 1979, 231). Exploitation by settled nomads and the peasantry, as well as serious problems of salination and drainage, necessitated agrarian reform and the organization of cooperatives and state farms. At the same time the need to utilize the water resources of the Euphrates received high priority.

A major dam on the Euphrates had been envisaged as early as 1927 by the French, but not realized. Shortly after independence in 1946, Sir Alexander Gibbs and Co. conducted a preliminary study for a dam near Yusuf Pasha which would have irrigated 100,000 ha. Nothing came of it, however, and this effort was followed by a twelve-volume study by the Soviets published in 1960. Next came a study by the West German Government in 1961 and another by the Dutch consulting firm, NEDECO, in 1963–1964 (Meliczek 1987, 111). The disruption of the U.A.R. and the breakdown of relations with the Germans in 1965 left the way open for Soviet participation in the building of the Tabqa Dam, which was officially inaugurated in July of 1973. The use of the waters of Lake Assad behind the dam has had a mixed history still to be resolved.

Proposed Irrigation

Table 8.1 outlines the proposed, revised, and actual irrigation projects relating to the Tabqa Dam. The Soviet proposal originally spoke of some 850,000 ha that could be irrigated with the waters of Lake Assad. This estimate was quickly downgraded by the Germans to 650,000 ha and then slightly revised by the Syrians to 640,000 ha. This total consisted of the six districts shown on Map 8.1 and below as well as in Table 8.2. Work relating to all these was unified and undertaken by the General Authority for the Development of the Euphrates Basin (GADEB).

A Pioneer, or Pilot, Project was initiated (Table 8.3) in May 1973 on the left bank of the Euphrates 18 km from the Tabqa Dam in the Raid area. The purpose of this project was to resettle nearly 60,000 villagers who had been flooded out by Lake Assad. Fifteen villages were built to replace the 59 that were abandoned along with the 31,231 ha of irrigated land and 7,495 ha of rain-fed land lost to reservoir flooding. It should also be noted that 82 percent of the displaced families owned sheep and 60 percent owned goats, the grazing land for which was in large part also lost (Meliczek 1987, 110). The original plan called for 19,600 ha to be irrigated, a figure which was to have been increased to 38,700 ha by the end

Table 8.1. Proposed, Revised, and Actual Irrigated Land Projects in the Syrian Euphrates Drainage Area
(all figures in hectares)

Location	Proposed Amt	Revised Amt	Actual Amount	Comments	Reference
Tabqa/Thawra	850,000 (Soviet est.)	650,000 (German est.)		See also LANDSAT reference sheet est. for private land	Bourgey, 346
		640,000 (Syrian decision)			Khayyat interview
		345,000 (1983)			
		135,000 (revised for 1980)		Deemed unrealistic	World Bank, 248
		40–60,000		Deemed more realistic	World Bank
		240,000 by 1980 "but by 1978 only 7,400 ha had been prepared"			USAID 1980 V.I, I-31
		"43,200 by 1980"			USAID 1980 V.I, I-31
Balikh (area #1)	185,000 200,000	185,000	— —		Bourgey, 346 Pitcher, 14
Euphrates Valley	240,000				Bourgey, 346

			1,600 11,500	"Central Euphrates Project"	Khayyat interview
—Lower Valley —(Area #2)	165,000 160,000	165,000	See Table 8.2 20,240 "Left bank near Raqqa"	"Underway 1974"	Sanlaville, 235 Pitcher, 14
—Lower Khabur —(Area #3)	70,000 75,000	75,000			Pitcher, 14 Sanlaville, 235
Rasafah (Area #4)	150,000 (Soviet est.)	20,000 (German est.)			Bourgey, 346 Sanlaville, 235
		25,000	none (1983)	Abandoned because of gypsiferous soil	Khayyat interview
Mayadin Plain (Area #5)	40,000	40,000			Pitcher, 14 Bourgey, 346
Maskanah-Aleppo (Area #6)	150,000 (125,000)	150,000			Khayyat interview (Ivanov, 77)
			15,000/13,282		Khayyat interview
—("near Aleppo")		(100,000)			(Ivanov, 77)

continued

Table 8.1. (continued)

Location	Proposed Amt	Revised Amt	Actual Amount	Comments	Reference
—northern and southern Aleppo region		180,000		Possibly recent addition to Area #6 in place of original lands	Khayyat interview
Khabur (upper)	400,000				Bourgey, 346
		137,900		See references this report	al-Ashhab, 41–42
Total: Areas 1–6		640,000		Original	
Eventual Total per Khayyat interview		345,000		Revised	
Eventual Total per Meliczek		240,000			
Totals including Revised Khabur Estimate		482,900 (or 377,900)			

Source: Bourgey (1979); Khayyat (1983); World Bank (1985); USAID (1980); Pitcher (1974); Sanlaville and Metral (1979); Ivanov (1983); al-Ashhab (1983); Meliczek (1987).

Of the 640,000 ha originally planned, 110,000 ha were to be irrigated by gravity flow from Lake Assad and 530,000 ha were to be irrigated by water pumped from the reservoir (Pitcher 1974, 14). Similar figures are cited by Meliczek (1987, 14): 120,000 ha and 520,000 ha.

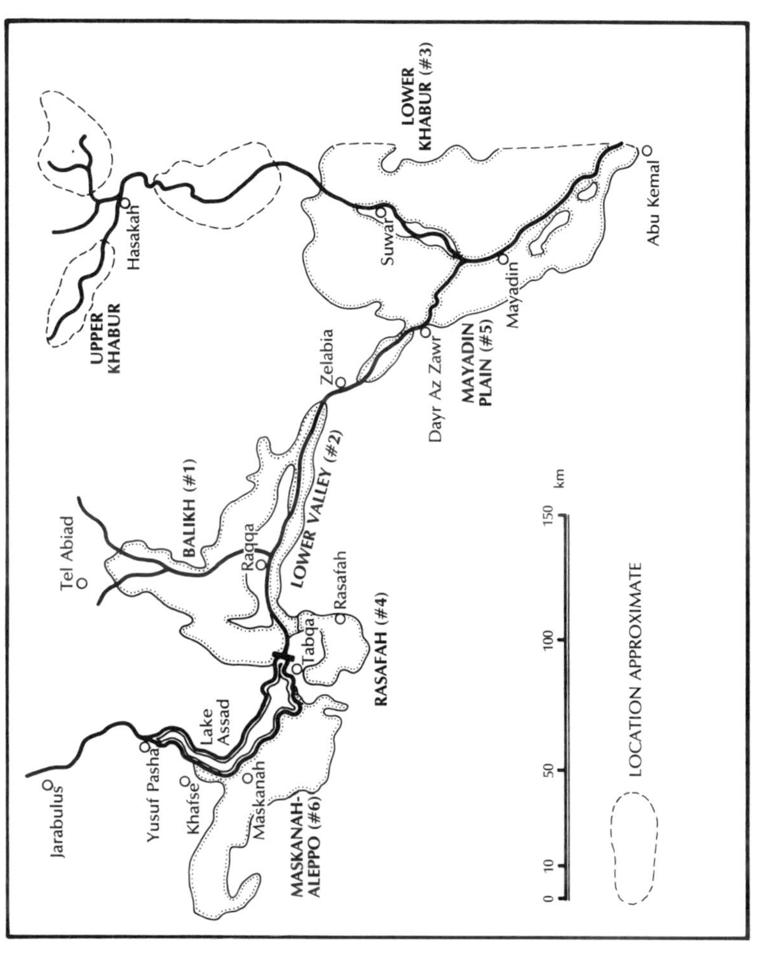

Map 8.1. Irrigation regions within the Euphrates River basin in Syria (Turkish border not shown). Source: Bourgey (1979, 347); *Syria Times* (Upper Khabur, 1985).

Table 8.2. Announced Quantities of Irrigated Land in the Syrian Euphrates Government Development Project Area, 1982–83 and 1985–86
(area in hectares)

District	Irrigable[1]	Irrigated 82–83[2]	Irrigated 85–86[1]	Total 82–83[2]	Status
Balikh					
Pilot	17,600	17,000	15,000	17,000	I
Zone 1	22,200	—	10,000	20,000	P
Zone 2	16,500	—	—	26,000	S
Zone 3	19,700	—	—		
Zones 4–7[a]	66,500	—	—	43,000	S
Subtotal	142,500	17,000	25,000	106,000	
Euphrates Valley					
Middle	23,000	12,000	12,500	12,000	I
		(10,000 net)		17,000	P
Lower[b]	—	—	—	50,000	S
Subtotal	23,000	12,000	12,500	79,000	
Meskene West[c]		24,300		24,300	I
		(21,000 net)		20,000	S
#21[d]	18,000		7,000		
#22[d]	20,000		20,000	30,000	P[e]
Subtotal	38,000	24,300	27,000	74,300	
TOTAL	203,000	53,300	64,500	259,300	

Source: 1 = Meliczek (1987); 2 = Metral (1987).

S = being studied   P = under construction   I = irrigated

Meskene East: Initiated but not completed as of 1986 (Meliczek 1987).

Rasafah/Mayadin/Lower Khabut areas abandoned or work suspended (Metral 1987).

[a] Includes Suwadiyah district.
[b] Initiated but not complete in 1986 (Meliczek 1987).
[c] Term used by Metral (1987).
[d] Terms used by Meliczek (1987).
[e] Includes Meskene East.

Table 8.3.   The Euphrates Valley Pilot/Pioneer Project

Plans prepared by Sir Alexander Gibb and partners: 1967

Begun: May 1973

Location: 18 km from Tabqa on the left bank of the Euphrates

Water: Served by Pump Station Kdeirane—6 pumps with a capacity of 25 m³ and a lift of 20 m.

Area:	Original	Third 5 Yr. Plan	Revised	Actual
	18,000	38,700*	32–19,000*	11,500
	(1973)*			(1983)
				15,000
				(1986)

Source: Bourgey (1979); Khayyat (1983); *Tishrin* (1985); Meliczek (1987).

*Proposed but not attained.
Note: This project was intended to resettle nearly 60,000 villagers who had been flooded out by the al-Assad Reservoir. Fifteen villages have been built replacing the original 59 that were abandoned.
The downward revision of the area cultivated was apparently the result of the large-scale collapse of the original canals and the loss of up to 30,000 m³/hr of water into the gypsiferous soils. While the canals have apparently been repaired, as recently as 29 July 1984 *Tishrin* reported "cracks" had appeared in the Balikh Canal.
The crops grown on the pilot project land were primarily cotton, but also barley, forage crops, sugar beets, corn, beans, fruit, and (for the first time) rice.

of the third Five-Year Plan. The crops to be grown were primarily cotton, but also barley, forage crops, sugar beets, corn, beans, fruit, and (for the first time) rice.

Revisions of Proposed Irrigation Goals

It was sometime after this that serious problems began to develop with regard to the application of water to the land. As summarized in the USAID 1980 report (2:II-1 and II-4) and intimated by various press releases from Syria, the Euphrates Basin soils are in large part gypsiferous, crusty, prone to erosion, and suitable only for careful applications of irrigation water. In a November 1982 interview with the press, Dr. 'Abduh Qasim, General Director of the

Public Establishment for Utilization of the Euphrates River Basin, spoke of the collapse of the canals leading to the Pioneer Project, when water was channeled through them, as well as of the loss of 5 m^3/s into the ground (Khayyat 1983). As recently as July 1984 (*Tishrin* 1985, 4) reported that "cracks" had appeared in the Balikh canal as a result of the gypsiferous soil on which it rests being dissolved by leaks from the canal.

USAID (1980), on page II-4, goes on to state that, "Class IV land is marginal at best for agriculture. Since only 64 percent of the land [in the Euphrates Project] is in classes I through IV, and 48 percent is Classes I through III, this suggests that less than half of the 640,000 ha is reasonably good land for irrigation purposes." This report then mentions and suggests a goal of 240,000 ha by 1980, but "by 1978 only 7,400 ha had been prepared," and suggests a projection for 1980 of 43,200 ha. In the interview cited above (Khayyat 1983) Dr. Qasim speaks of the possibility of up to 345,000 ha being irrigated eventually. Meliczek (1987, 129) reports that by 1987 an eventual goal of 240,000 ha was to be irrigated by GADEB.

To these figures should be added the lands of the upper Khabur basin, which will also receive irrigation water. These were originally estimated to be 400,000 ha but a recent news release (al-Ashhab 1983) gives a total of 137,900 ha for three sub-projects (Table 8.4).

Thus it would seem realistic to anticipate water being applied from Lake Assad to an absolute maximum of 345,000 ha or, more realistically, 240,000 ha plus another 137,900 ha on the Khabur. This is not the entire story, however, and the details follow.

In Rasafah (area 4) the Soviets suggested 150,000 ha; the Germans proposed 20,000 ha because of the gypsiferous soils; and the Syrians apparently planned on 25,000 ha. Qasim indicates in his interview (Khayyat 1983) that the entire project has been abandoned. He also mentions that, while large tracts of the original Maskanah-Aleppo district have been withdrawn from possible irrigation, new lands in the northern and southern Aleppo region totalling 180,000 ha are to be added. (These

Table 8.4. Dams in the Euphrates River Basin in Syria
(including the Khabur River)

Dam/Project Name	Storage Capacity	Reservoir Area	Comments
Tishreen (1.6 MW)	1,300 Mm³	70 sq km	planned, *MEED* 1986a
Tabqa (800 MW)	11,600 Mm³	625 sq km	see text and tables
Baath (64 MW)	90 Mm³	2.7 sq km	completed 1986, *MEED* 1986
Western Hasakah[a]	91 Mm³	1,020 ha	
Eastern Hasakah[a]	232 Mm³	3,100 ha	49,450 ha combined
Khabur[a]	665 Mm³	9,580 ha	46,450 ha
Diversions, Ras al-Ayn Springs to Khabur			42,000 ha
		TOTAL	137,900 ha

Source: al-Ashhab (1983).

[a] Under construction March 1983.

Dam/Project Name	Storage Capacity	Reservoir Area	Comments
Bab el-Hadeed/ al-Jawayda	—	—	2,800 ha combined
al-Jarah	23 Mm³	—	—
Mashouq	2.5 Mm³	—	300 ha
Jagh Jagh	—	—	1,200 ha
Malkeva	61 Mm³	—	600 ha
al-Hakima/ al-Mansouria	1 Mm³	—	400 ha combined
		TOTAL	5,300 ha

Source: Sati (1982).

Dam/Project Name	Storage Capacity	Reservoir Area
al-Wa'ar (Deir ez-Zor)	3.345 Mm³	805,000 sq m
Karima (Hasakah)	1.9 Mm³	800,000 sq m
Abou al-Kahef (Raqqa)	0.62 Mm³	390,000 sq m

Source: SAR (1980, 68, Table 8.1)

changes are apparently taken into account in the total quoted in the above paragraph.)

Production Achieved by State-Run Projects

There remains the question of just how much land has actually been prepared and how much is actually being cultivated. No data later than 1986 are available, and comments based on those may need upgrading. Nevertheless, the actual amount of land successfully brought into production seems small. Qasim gives 13,100 ha for a "Central Euphrates Project," presumably part of the Euphrates Valley Project previously mentioned by Qasim (Khayyat 1983), while Mileczek (1987) limits this figure to 12,500 ha in 1986 plus an additional 10,000 ha in area 1 of the Balikh.

The Pioneer Project was revised downward to 32,000 and then 19,000 ha, although in 1983 only 11,500 ha were cultivated. Subsequently, Mileczek (1987) reports 15,000 ha irrigated and farmed on the Pilot Project in 1986. Another 27,000 ha in the Maskanah-Aleppo area round out this accounting to a 1986 total of 64,500 ha irrigated.

The slow progress being made can be appreciated by contrasting the status report on the Euphrates River Irrigation Project for 1976 (U.S. Department of State 1976) described in Table 8.5 with the amounts given above and in Tables 8.1 and 8.2. In 1976 USAID listed 95,000 ha as developed, designs completed, bids invited, and contracts signed for the Pilot Project, Balikh, and the Mid-Euphrates areas. By 1985–86 only 37,500 ha were actually being irrigated in the same areas (Mileczek 1987, 126). Nevertheless, the project is being pushed forward according to Metral (1987, 119), who lists 89,000 ha being studied in detail, 117,000 ha work in progress, and 53,300 ha irrigated—a total of 185,000 ha as of 1983.

However, these figures include the Maskanah area. Without their inclusion, figures comparable to USAID's would be: 69,000 ha being studied, 37,000 ha work in progress (excluding 50,000 ha in the lower valley), and 29,000 ha irrigated—a total of 135,000 ha. There appears to be an overall increase of 40,000 ha involved in all aspects of development, but a decline in actual irrigated area

Table 8.5. Status Report on Euphrates River Irrigation Project

Date: July 22, 1976:		Area ha
Pilot Project	developed	20,000
Balikh (sect 1)	construction contracts signed	10,000
Balikh remaining	bids invited	12,000
Balikh (sect 2)	designs completed	26,000
Mid-Euphrates Valley	construction contracts signed	27,000
	Total ha:	95,000
Main and branch canals	800 km	
Secondary canals and flumes	900 km	
Main drains (surface)	500 km	

Source: USDS (ca 1976).

between 1983 and 1986 of 8,500 ha. Meliczek's figure of 165,500 ha "considered irrigable" for the area in 1986 is 20,000 ha less than Metral's for 1983.

Can such confusion and shortfall be possible? When one reads the Qasim/Khayyat interview (Khayyat 1983) in full, the litany of bureaucratic ineptitude, engineering over-optimism, and the true difficulty of the region itself make this track record seem within reason. Another indicator of the seriousness of production problems in the Euphrates Valley is the call for bids for work on drainage systems by the Irrigation Ministry (*MEED* 1984a, 33). This same article mentions a report made by the French consortium of Gersar and SCET International which found about 3,000 ha/yr being affected by salinity and poor drainage. Add to this the 31,231 ha lost when Lake Assad was formed (Meliczek 1987, 110), and the lack of results comes into focus.[1] Further evidence of the

1. The impact of uncontrolled pumping on groundwater in Syria as well as the use of groundwater drawn by the Turks from aquifer recharge areas in Turkey will have a profound affect on this resource. This topic is discussed in the section of this report dealing with the upper Khabur and Ceylanpinar areas.

disappointments in the Euphrates Project is that the latest Five-Year Plan for Syria emphasizes rain-fed agriculture and the semi-arid steppe lands. On the other hand, it should be kept in mind that large tracts of land are being irrigated and cultivated by private farmers large and small.

Privately Cultivated Land

Privately cultivated land is the major consumer of Euphrates water in Syria. As with other data, statistics relating to the exact amount are sparse, incomplete, and seldom current. There are two main sources of these data. The Syrian government releases figures from time to time; these have been available to this writer largely through references in secondary sources. Another group of data comes from LANDSAT imagery and an evaluation of "intensively cultivated" and other categories of land included in the USAID (1980) report. By their definition, "intensively cultivated land" is considered to be irrigated.

The problem with the latter data, aside from technical difficulties always associated with imagery interpretation, is that the report uses a series of land classifications which are discontinuous in space. That is, the areal units used to define and aggregate information may occur in two or more widely separated places with only cursory indications of what is found within subunits. Syria has been divided into 58 Resource Planning Units (RPUs) by the USAID report (1980, see Map 8.2); each RPU in turn consists of several Production Planning Areas (PPA). Discriminating among PPAs in a given RPU can seldom be exact. Table 8.6 shows the amount of irrigated land in selected regions of northern Syria as reported from several sources. In this case, general geographic and/ or political subunits are the basis for reporting. Table 9.7 relies upon LANDSAT data presented in table form elsewhere in the USAID report. Map 9.2 shows the RPUs for northern Syria. The discontinuous character of units 31, 32, 40 and 57 should be noted.

Given the above caveats, the following may be stated. Treakle (1970) reported as of 1970 that 160,000 ha of irrigated land were found in the Euphrates valley. This was clearly before Lake Assad

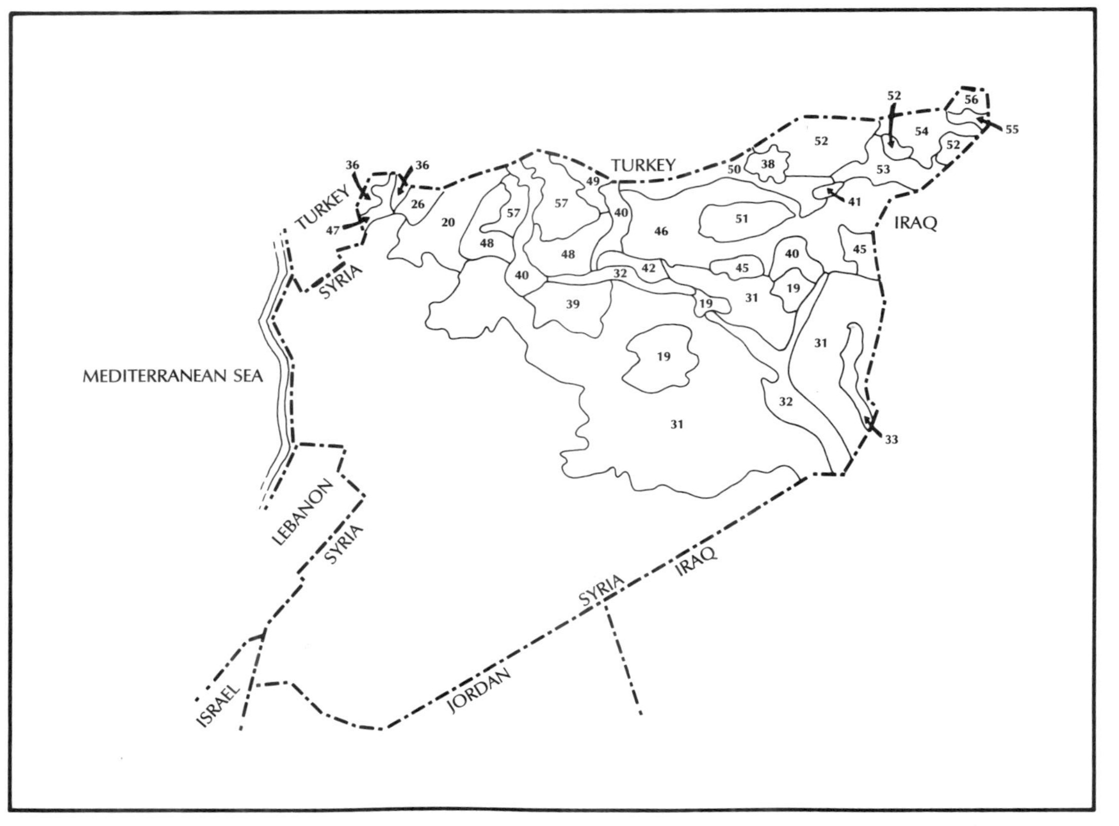

Map 8.2. Resource planning units of northeastern Syria. Source: USAID (1980, I).

Table 8.6. USAID/SAR Estimates of "Intensively Cultivated Land" in Selected Regions of Northern Syria
(includes LANDSAT imagery)

Location	Amount in ha	Comments	Source
Euphrates Valley	160,000	as of 1970	Treakle, 9
	(−28,000)	flooded by Lake Assad	Samman, 23
	(−25,000)	flooded by Lake Assad	Pitcher, 15
	(−31,231)		Meliczek, 110
Lower Euphrates	142,000	Private, to be integrated into project	USAID 1980 [RPU 32], V.2, i–111
	(128,000)	(1974)	Meliczek, 116
Raqqa Mohafaza	60,773	LANDSAT	USAID, V.3, 1–85
Deir ez-Zor Mohafaza	85,676	LANDSAT	USAID, V.3, 1–87
TOTAL a:	146,449		
Hasakah Mohafaza	80,909	LANDSAT	USAID, V.3, 1–82
"around Hasakah"	25,000 approx.	location unclear, RPU 50	USAID, V.2, 1–163
	(4,542)	("irrigation network #2")	
"Upper Khabur"	60,000	"irrigation network #3", RPU 40	USAID, V.2, 1–137
TOTAL b:	85,000	Total, drawn from Syrian sources, approximates LANDSAT data.	

TOTAL a + TOTAL b = 231,449 ha

Sources: Treakle (1970); Samman (1980); Pitcher (1974); USAID (1980); SAR (1980; Meliczek (1987).

Note: "Intensively cultivated land" by USAID definition is considered to be irrigated; in the cases cited here water would come from surface sources.

was filled. Samman (1980) and Pitcher (1974) both report about 25,000 ha of land lost due to flooding, while Meliczek (1987, 110) cites 31,231 ha. In 1974, S. El Abd conducted a study of the valley areas where farmers displaced by reservoir flooding would be relocated. His survey indicated 128,000 ha of irrigated land downstream from the Tabqa Dam (Meliczek 1987, 116). This amount plus the lost hectarage mentioned above comes to 159,231 ha, almost identical with Treakle's figure. USAID (1980) in 1976 observed/estimated 142,000 ha of land irrigated in the "lower Euphrates." These latter LANDSAT data are slightly larger than El Abd's and Treakle's figures given losses from flooding and perhaps some increase in irrigation along the edges of the reservoir and elsewhere downstream.

A cross-check on these figures comes about when irrigated land in Raqqa Mohafaza (60,773 ha) is combined with that in Deir ez-Zor Mohafaza (85,676 ha), giving a total irrigated land downstream from Tabqa of 146,449 ha—close to the 142,000 ha cited above (Table 8.6). While both of these sources come from USAID (1980), the slightly smaller figure apparently is derived from Syrian sources, while the larger is the result of LANDSAT analysis.

In the same way, two corroborating figures are given in USAID (1980) for the Khabur tributary. Hasakah Mohafaza is listed as having 80,909 ha of irrigated land, while areas "around Al-Hasakah and in the Upper Khabur" are listed as having "approximately 25,000 ha" and 60,000 ha respectively.

Table 8.7 allows a slightly different view of the situation, but with approximately the same results. RPUs 32, 40, and 42 essentially comprise the valley of the Euphrates River. Two additional parcels of RPU 40 are found along the Balikh and west of the lower Khabur. Little irrigated land is currently found in the latter unit; it would appear that most of the 50,000 ha attributed to this RPU are in the basin of the Balikh. In any event, water use and depletion from such fields will decrease downstream discharge of the main stream. RPU 42 is in the Al-Raqqa area, while RPU 32 would represent the Deir ez-Zor area as well as part of the lower Khabur and an area downstream from the Tabqa Dam.

RPUs 50, 38 and 41 cover most of the upper Khabur system

Table 8.7. Intensive Agriculture: Northeast Syria[a] as Determined from LANDSAT, 28 July 1976
(areas in hectares)

Euphrates, Balikh, Lower Khabur (Pumped River)		Khabur Tribs. (Pumped River, Groundwater)		Partial Euphrates Drainage (Prob. Pumped River)		Within Euphrates Basin (Probably Groundwater)		Queik System		Within Basin (No Intensive Agriculture)	
RPU	Area	RPU	Area	RPU	Area	RPU	Area	RPU	Area	RPU	Area
32	145,000	50	24,500	19	1,700	39	7,100	20	31,200	33	—
40	50,000	38[b]	3,400	31	7,240	45	200			51	—
42	9,200	41[b]	2,500			46	27,300			53	—
						48	4,600			54	—
						49	12,000				
						50	200				
						57	19,800				
TOTAL	201,700		30,400		8,940		71,200		31,200		—

Total of first three columns = 241,040 ha

Source: USAID (1980, I-210, Table 3.)

Note: See Map 8.2 for location of RPUs.
[a] Tigris drainage excluded.
[b] Partially within basin but all irrigation included.

Table 8.8. On-Line Government Project Lands, ca 1980

Balikh	21,200 ha
Central Euphrates Project	13,100 ha
Maskanah	13,282 ha
Total	47,582 ha

and as such also diminish downstream flow. These six units in sum account for 232,100 ha. Combined with the 8,940 ha in units 19 and 31 (which in all likelihood receive pumped water from the Euphrates and Khabur), the 241,040 ha thus noted are close to the 231,449 ha listed in Table 8.6.

RPUs 39, 45, 46, 48, 49, 50, and 57 are more difficult to assign to river flow or groundwater use. The latter is probably more the case and will be treated again in the Chapter 9 relating to the Khabur.

The Queik River, while outside the Euphrates drainage, is mentioned for two reasons. Although previously the source of water for Aleppo, its waters are no longer sufficient for that purpose, in large part because of upstream diversions in Turkey and Syria. As a result, the city of Aleppo now depends upon Euphrates waters pumped from Lake Assad. Current use of 80.3 Mm^3/yr is considered inadequate, and this city's dependency upon the Euphrates must continue and grow. (See Chapter 6.)

The remaining RPUs—33, 51, 53, 54—while within the study area show no intensive agriculture and in part fall outside the drainage basin.

The 142,000 ha cited above for the Euphrates valley plus an additional 81,000 ha on the Khabur River made a total (*circa* 1980) of 243,000 ha irrigated. Included within this LANDSAT-based total would be 47,582 ha apparently on-line through government sponsored projects (see Table 8.8) or the 64,500 ha of government irrigated land cited by Meliczek (1987) for 1986. In either case, it is not clear exactly how much of this land was irrigated previously by private holders. Additional data are shown in Tables 8.9 and 8.10. Some confusion results form a lack of explanation of the terms (and the overlap of areas) "winter crops" and "summer crops."

Even assuming that much double cropping is practiced, water would still have to be applied twice to the same parcel of land in

Table 8.9. Water-Fund Depletion Resulting from Evapotranspiration and Related Deficits

Location	Area Irrigated 1000 ha	Deficit Replacement m³/ha	Fund Depletion m³/ha	Total System Depletion m³	Amt Returned to System m³/ha	Total Returned to System m³
Lower Euph. (Deir ez-Zor)	345 (240)[a]	10,360	16,835	5808 (4042)	9,065	3127 (2176)
Ras al-Ayn (Upper Khabur)	42	7,070	11,489	483	6,186	260
Tel Tamer (Hasakah)	95.9	7,720	12,545	1203	6,755	648
Totals	482.9 (377.9)	—	—	7486 (5728)	—	4035 (3084)
Private Lands (as of approx. 1980) per LANDSAT Imagery (May include government sponsored irrigation—see below)						
RPUs Euphrates						
32, 42	151.7	10,360	16,835	2554	9,065	1375
40	50.0	7,070	11,489	574	6,186	309

Khabur						
50	24.5	7,070	11,489	281	6,186	152
38, 41	5.9	7,720	12,545	74	6,755	40
19, 31	8.94	10,360	16,835	151	9,065	81
Totals	241.040	—	—	3634	—	1957
Government Sponsored Irrigation[b] circa 1980						
	47.6[c]	7,720	12,545	597	6,755	321

Source: See Table 8.4 for supporting materials and discussion.

[a] Figures in parentheses show amounts based on 240,000 ha total (Meliczek 1987).
[b] Probably included in LANDSAT totals given below.
[c] See Table 8.2 for areas irrigated in 1982 and 1986.

Table 8.10. Irrigated Land in the Euphrates Drainage Basin, Northern
Syria, 1979–81
(values in hectares)

Year	Mohafazat	Winter Crops	Summer Crops	Fruit Trees	Total	Year Total
1979	al-Hassakeh	49,126	42,412	1893	93,431	
	al-Raqqa	27,823	26,341	136	54,300	
	Deir ez-Zor	45,892	59,455	3406	108,753	256,484
1980	al-Hassakeh	45,820	44,217	2218	92,255	
	al-Raqqa	20,981	22,883	43	44,001	
	Deir ez-Zor	47,455	72,584	3710	123,749	260,005
1981	al-Hassakeh	41,762	42,829	2325	86,916	
	al-Raqqa	20,829	18,853	186	39,868	
	Deir ez-Zor	48,454	70,134	4022	122,610	249,394

Source: SCBS (1979; 1980; 1981, Table 9).

that case. Both tables show variation from year to year which falls
within a reasonable range. The greatest difference comes between
yearly totals for the two tables. No immediate explanation of such
variation is forthcoming, but this may be explained if one set of
data comes from canal gauges and the other from aerial or other
surveys. In any event, the average of all five values given on these
two charts is 240,711 ha. This is for all practical purposes the
same as the 243,000 ha cited above, a figure arrived at through
completely different sources.

If a conservative figure of 241,000 ha is accepted and if either
the 47,500 or 64,500 ha of government land are deducted (see
above), then independent farming should come to about 196,500
or 176,500 ha of privately irrigated land, although the absolute
amount of land receiving irrigation water would remain the same.
In any event, the above total represents recent usage and should
be close to what was being consumed in 1986.

Water Depletion from Syrian Irrigation on the Euphrates

The method by which depletion of river water through evapotranspiration and system inefficiency is computed was presented for

Table 8.11. Sources of Irrigation Water in the Euphrates Drainage Basin, Northern Syria, 1980–81
(values in hectares)

Year	Mohafazat	Pumped from Wells	Pumped or Free Flow from Rivers	Total	Year Total
1980	al-Hassakeh	33,479	49,164	82,643	
	al-Raqqa	16,098	34,067	50,165	
	Deir ez-Zor	1,170	79,548	80,718	213,526
1981	al-Hassakeh	34,828	52,413	87,241	
	al-Raqqa	17,698	41,779	59,477	
	Deir ez-Zor	1,170	76,260	77,430	224,148

Source: SCBS (1980; 1981, Table 6).

Turkey with best estimates of such demands given in Table 7.4. A similar presentation for Syria is now possible using the values already derived and with reference to the amounts of irrigated land discussed above.

Table 8.11 presents two sets of values. The first is based on the revised plans for irrigating Syrian lands with Euphrates waters. The second presents best estimates for the actual amount of water removed from the system on or about 1980.[2] As mentioned above, data are lacking for more recent periods, but the slow addition of new irrigated lands, the probable loss of land through salination and drainage problems, and the substitution of government sponsored irrigation projects in areas previously privately farmed mean that the amount under actual production today is likely to approximate the amounts shown in this table.

In summary, 241,000 ha of private and government lands require about 3,600 Mm^3/yr of water. An estimated return flow of about 2,000 Mm^3 (making a total withdrawal of approximately

2. While the PE values given in Tables 8.4 and 9.7 have been calculated, the FAO report on the Khabur (79–80) gives two similar empirical values. Cotton in the Khabur area requires 120 days (15 May to 1 Oct) and 10,000 to 12,000 m^3 of water/ha. (This would not include losses due to system inefficiency.) Another study showed that 17,700 ha cotton, 2,200 ha fruit and legumes, and 4,400 ha cereals used 240×160 m^3 of water, or approximately 1 m^3/m^2. These examples are in essential agreement with the values used for the computations described here.

5,600 Mm^3) while augmenting stream flow cannot help but increase downstream salinity.

If the full 345,000 ha suggested by Qasim for the Euphrates are realized along with another 137,900 ha on the Khabur, water depleted from the system will double as will return flow. In order to fully evaluate the impact of these volumes upon the total Euphrates system, upstream uses in Turkey must be considered along with another major source of water loss, evaporation from reservoirs and canals. The special case of the Khabur with its source areas in Turkey also must be considered before turning to a final accounting of Euphrates waters in both countries.

9

The Khabur River and Its Tributaries

Syria north and east of the Euphrates River is drained by the Balikh and Khabur River systems. These streams enter the Euphrates from the left bank below the Tabqa Dam and provide on the average 0.6 percent and 6 percent of the total flow of the river respectively (Table 6.1). While this amount is relatively small, the significance of these tributaries is disproportionately great, particularly in the case of the Khabur. The reasons for this are threefold. Syrian efforts at agricultural development have met with numerous frustrations along the mainstream of the Euphrates, while the lands of the upper Khabur offer promise of success. The Khabur is cited as Syria's significant contribution to the discharge of the Euphrates and offers a *quid pro quo* basis for Syrian claims to use of the river. Discharge from these tributaries significantly affects the amount and quality of water passing into Iraq.

Evidence will be presented that more than 80 percent of the waters of the Khabur and its tributaries originate in Turkey and can and will be affected by that country's development plans. This, in turn, will affect Syria's plans for the area as well as the third riparian user, Iraq.

This region is known in Syria as the Jezirah and is further divided into the Lower Jezirah, which stretches north from Deir ez-Zor on the Euphrates to the Jebel Abd el-Aziz on the west and the Jebel Sinjar (mountains) on the east of the Khabur River, and, north of this barrier, the High Jezirah, which extends from Hasakah in Syria at the confluence of the Khabur and Jagh Jagh rivers to the Anti-Taurus Mountains in Turkey. This gently rolling plain is the catchment area for the waters of the Khabur system which lies 45 percent within Turkey (10,722 sq km) and 55 percent within Syria (13,575 sq km). Another, approximately 1,600 sq km, falls within the borders of Iraq to the southeast. However, this area as open desert contributes nothing to stream flow. Rainfall in the Lower Jezirah is less than 300 mm/yr and near Deir ez-Zor evapotranspiration (1,504 mm/yr) is more than ten times annual precipitation (148 mm) (Map 9.1). Elevations as well as rainfall increase steadily to the north (Table 9.1).

The highest elevations in the upper basin of the Khabur are 1,919 m at Karacali Dag (mountain); near Mardin, Turkey (1,200 m); and in the south the Jebel Abd el-Aziz (920 m) and the Jebel Sinjar (1,460 m). The course of the Khabur River extends for approximately 120 km in Turkey with a slope varying between 5.2°/00 and 31°/00. It flows for another 486 km in Syria to its confluence with the Euphrates at Bseira near Deir ez-Zor. In Syria its descent is much more gradual, ranging from 0.27°/00 to 0.5°/00. Near Ras al-Ayn the valley of the Khabur is 2 to 4 km wide, while south of Suwar it flows across a desert plain. A number of tributaries enter the Khabur from its left bank. Among these, the Djirdjib, the Zergane, and the Jagh Jagh would be permanent streams save for summer depletions of irrigation water. Others, the Breibitch, the Jarrah, Khneizir, and the Roumelie flow only during the height of the rainy season (for instance, in 1963 they had gone dry by July). The disposition of these streams is shown on Maps 9.2, 9.3, and 9.4.

A main feature of the eastern Jezirah is the Radd Marsh formed by the uplift of the Jebel Sinjar in the late Quaternary. This blocking of the south-flowing streams diverted them westward to the Jagh Jagh. Evapotranspiration in the Radd is so great, however, that only in times of flood does water find its way in any quantity west to the Khabur.

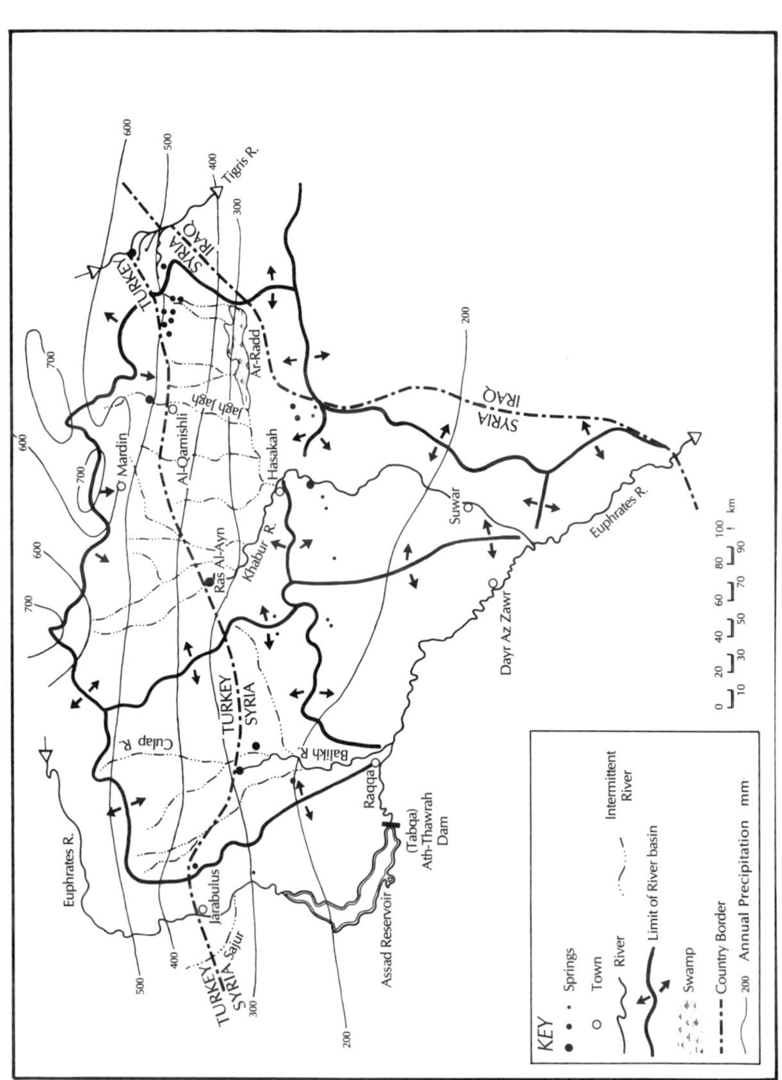

Map 9.1. Precipitation of northeastern Syria (including its extensions in Turkey). Source: FAO (1966, Map 1 and endpapers).

Table 9.1. Locations, Elevations, and Precipitation in Syria and Turkey

Location	Elevation in meters	Annual Avg. Precip. mm FAO	GAP	Years Record	Period of Observation FAO Data
Mardin	1150	686	714	(39)	1930–1960
Siverek	850	546	548	(48)	1930–1959
Gaziantep	840	550	555	(46)	1930–1959
Diyarbakir	677	488	488	(49)	1930–1959
Viransehir	575	537	540	(27)	1930–1959
Nusaybin /Qamishli	500	463*	485	(25)	1954–1960
Urfa	547	452	470	(46)	1930–1960
Qamishli /Nusaybin	467	452*	485	(25)	1952–1960
Ras al-Ayn /Ceylanpinar	350	292*	333	(23)	1957–1961
Tel Tamer	335	309			1948–1961
Hasakah	300	267			1931–1960
Raqqa	251	174*			1953–1950
Deir ez-Zor	200	148*			1931–1960
Abu Kemal	174	100*			1959–1960
Khafsa	350	201			1957–1960
Maskanah	350	201			1957–1960
Jarabulus	350	331			1949–1960

Source: FAO (1966, Table III-1); GAP (1980, Table III-1).

*Adjusted by FAO to reflect long-term projections.

Hydrogeology of the High Jezirah

The High Jezirah is bounded structurally on the north by the Mardin anticline and fault line. To the south the anticline and uplift of the Jebel Abd el-Aziz disrupts the stratigraphic continuity of the region. Within these limits are a series of south-dipping strata ranging in age from the Middle Cretaceous to the Quaternary and Pliocene. These beds are of great importance for among them are aquifers which provide the overwhelming share of water found in the Khabur and its tributaries (Figure 9.1).

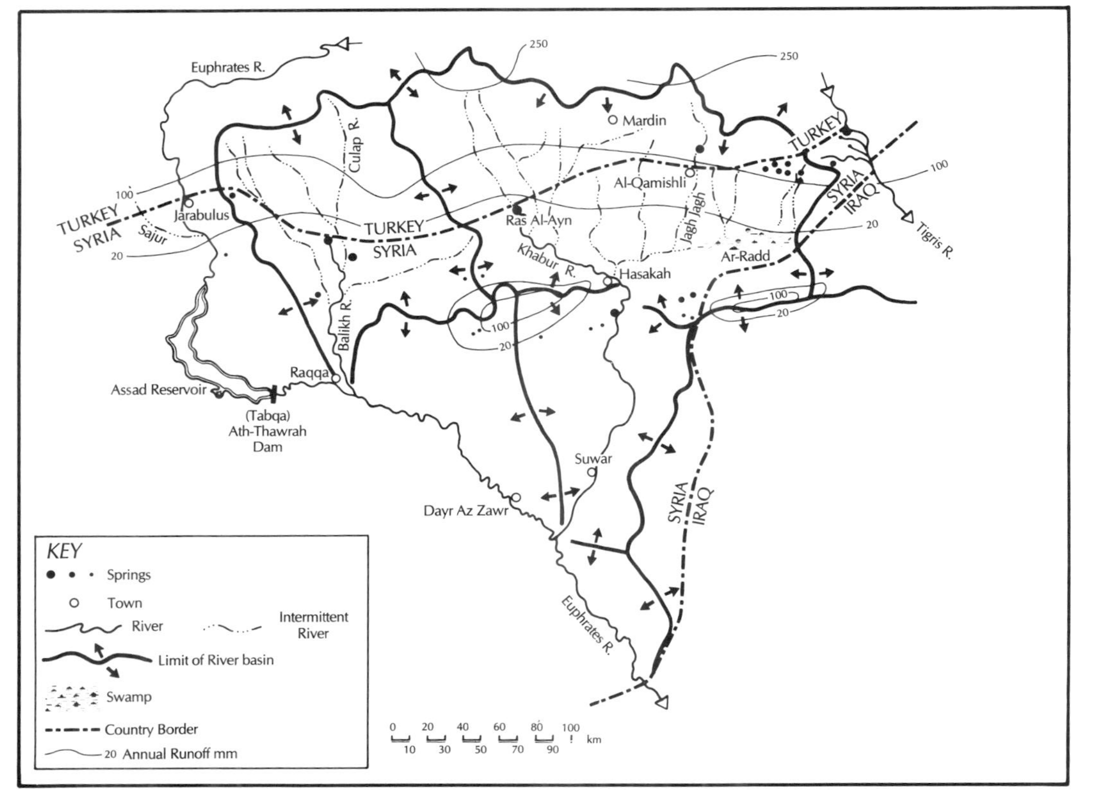

Map 9.2. Hydrography of northeastern Syria (including its extensions in Turkey). Source: FAO (1966, Figure II-4).

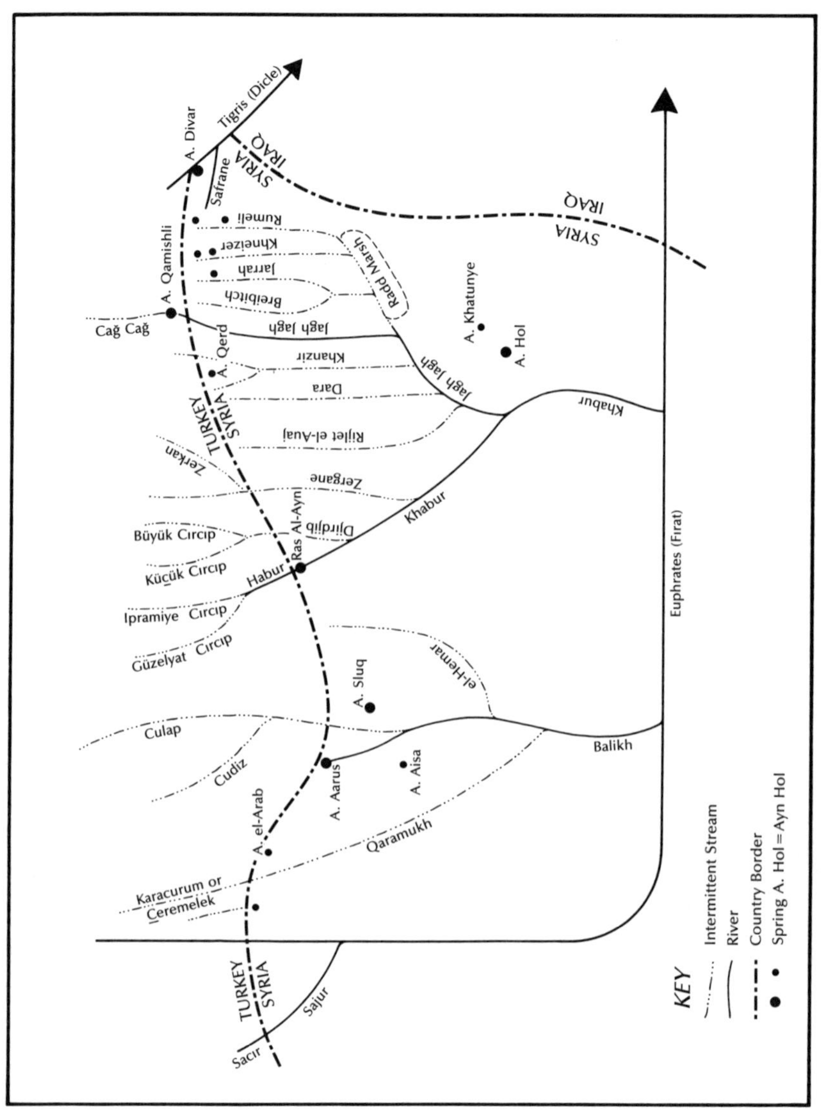

Map 9.3. Streams and springs of the Jezirah.

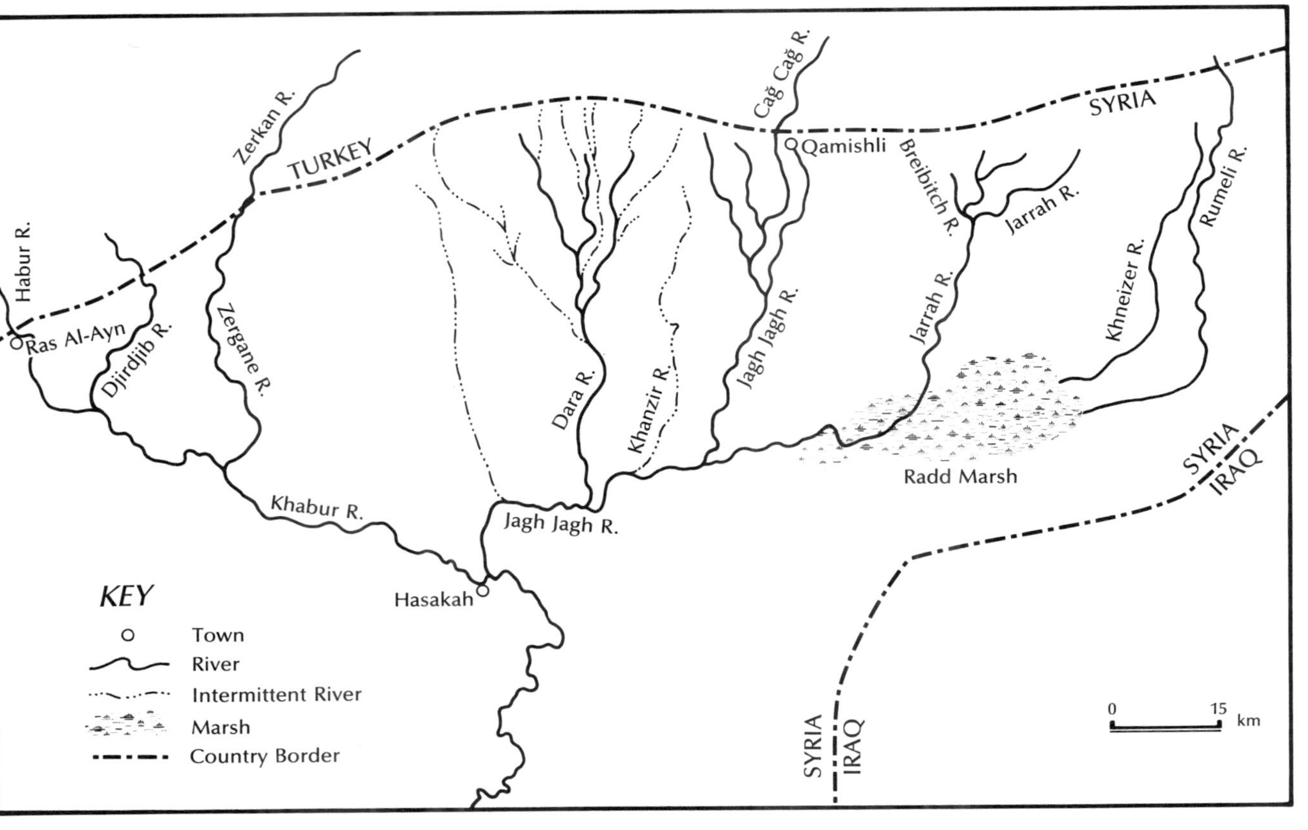

Map 9.4. The Khabur River and its tributaries.

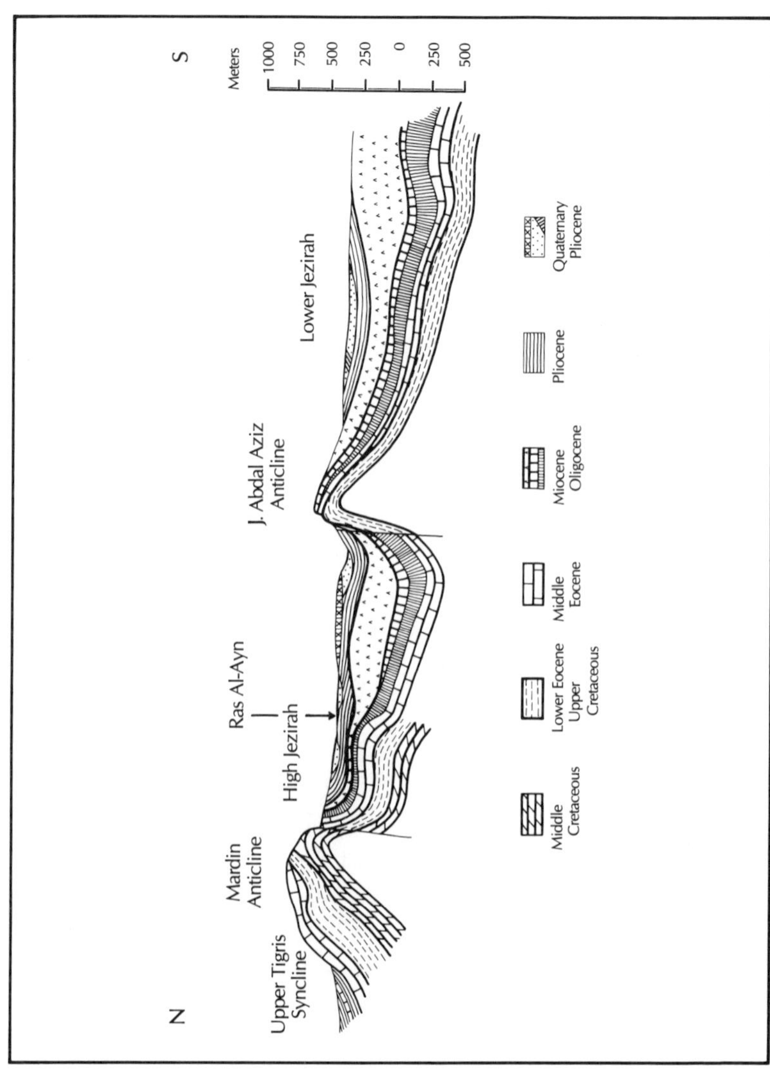

Figure 9.1. Schematic north-south geologic cross-section of the Jezirah. Source: Based on FAO (1966, Figure IV-8).

Four distinct assemblages of strata have been identified which constitute the major aquifers of the Jezirah (Map 9.5).

1. *Eocene/Oligocene limestones and dolomites*: These strata, where they are exposed to the north in Turkey, serve as the principal recharge area and subsequently form the major aquifer providing water for the Ras al-Ayn and other Syrian springs (Tables 9.2, 9.3). They have numerous open passageways for direct flow as well as great fissured storage capacity. It is estimated that, of the 2 billion m^3 of water supplied to the catchment area in Turkey by precipitation each year, perhaps 400 Mm^3 consist of runoff while the remaining 1,600 Mm^3 recharges this aquifer. The major exfluents of all this are the Ras al-Ayn and the Ayn Aarus near Tel Abiad on the Balikh. Even more impressive are the subterranean reserves, which account for the steady and nearly unvarying flow of these springs. A minimum of at least eight times the annual volume of flow would account for such regularity. The quality of the water thus delivered is good, with some exceptions where sulphur content makes it less acceptable for agriculture. Of the more than ten springs making up the Ras al-Ayn, two are named Ayn Kibrit (the Spring of the Match) indicating the presence of sulphur.

2. *Gypsiferous and calcareous rock of the Middle and Upper Miocene*: Less porous and permeable than the strata described above, these beds have varying capacities as aquifers with the best occurring where fissuring due to tectonism has taken place. The exposure of these beds largely near the Jebel Abd el-Aziz in an area of greatly reduced precipitation also limits both their recharge capacity and the total amount of water which they provide. A total flow of 2 to 3 m^3/s, of which 1 to 2 m^3/s surfaces as springs and the remainder as evaporation, limits the effectiveness of this source. Furthermore, karst solution in the gypsum makes the quality of the water highly variable.

3. *Argellites of the Pontico-Pliocene*: While these rocks are not entirely impermeable, they provide little opportunity for storing large amounts of water. An estimated total flow of 0.5 m^3/s and poor quality characterizes these waters.

4. *Pliocene-Quaternary unconsolidated materials*: These sands, sandstones, gravels, conglomerates, and basalts have excel-

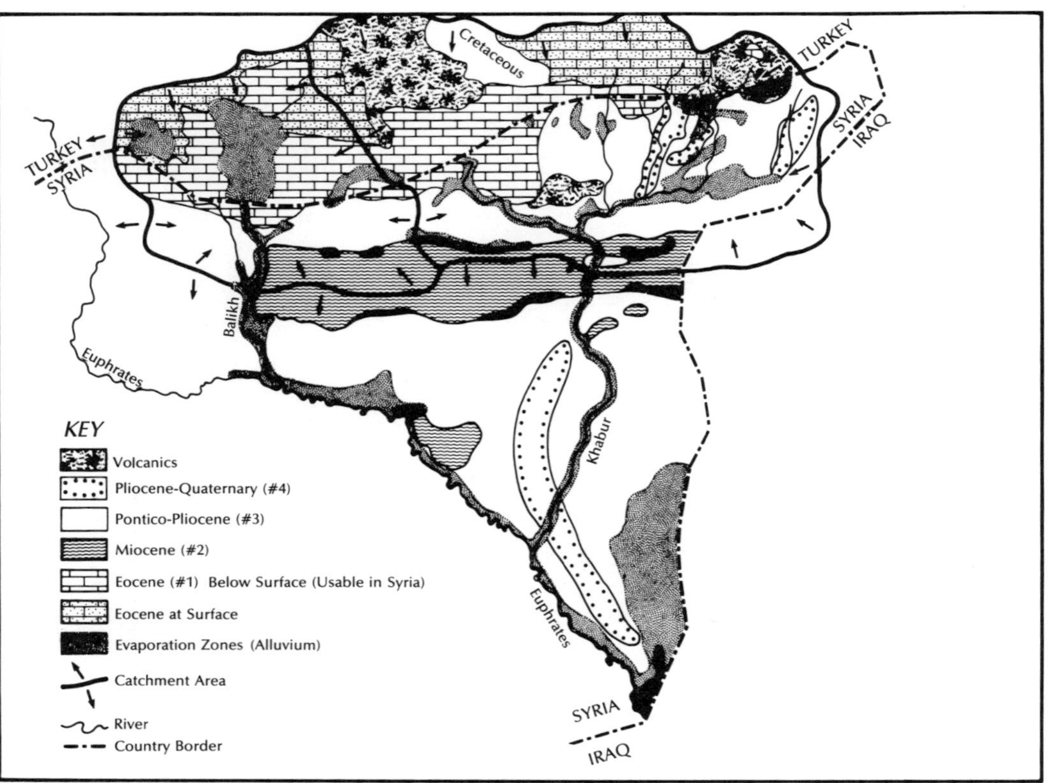

Map 9.5. Aquifers in the catchment area of the Jezirah. Source: FAO (1980); Tanoglu et al. (1961, plate 4).

Table 9.2. The Ras al-Ayn (Springs)

Name	Altitude m	Flow in m^3/s	Date
A. Hassan (south)	345.3	2.73	15/4/60
		1.86	11/8/60
A. Kibrit (south)	344.3	4.16	22/4/60
		4.16	14/8/60
		5.52	21/3/60
A. Zerga (south)	344.3	5.15	15/4/60
		6.35	11/8/60
Number 2 (North)	344.5	3.11	15/4/60
Number 7 (Zerga N.)	347.5	0.42	23/4/60
The Khabur River (100 m downstream from the frontier)	344.5	1.93	2/8/60
The Khabur River (Right branch upstream of the confluence)		21.6	2/8/60
		20.7	9/8/60
The Khabur River (Left branch 350 m upstream of the confluence)	344.3	21.4	3/8/60
		20.8	8/8/60
The Khabur River (Downstream from the confluence two branches.)	344.1	40.7	4/8/60
		41.6	4/8/60
		41.0	9/8/60

Names of springs:	Left Branch:	Arkhum	Right Branch:	Halaf
		Zerkan		Hassan
		Djamus		Jabbar
		Banos		Zerga
				Kibrit-1
	Main stream:			Kibrit-2

Source: FAO (1966, Figure II-7 and Table II-1).

Table 9.3. Springs of the High Jezirah

Northeast

More than 100 springs; most with flow less than 0.0025 m³/s.
Water quality is excellent.
Temperature less than 18° C.
Residue less than 0.5 g/l. (.0005 g/m³)

	Flow
Ain Divar	0.015 m³/s
Hanauye	0.012 m³/s
Baba Sinar	0.032 m³/s
Der Guessen	0.030 m³/s

Mid-Central

Approximately 35 springs; highly variable flow.
Grouped around the Jebel Abd al-Aziz.
Water quality varies; that from limestones is good.
Temperature: 19°/24°C.
Residue: 2.7/26* g/l. (.0027/.026 g/m³
Ayn Jibissa

	Flow
Lake Khaturye	0.500 m³/s
Ayn Hol	0.300 m³/s
Tel Tabane	0.600 m³/s
Ain Aissa	0.050 m³/s
Um Madfa	0.030 m³/s

Northern Frontier

	Flow
Ayn al-Arab	0.150 m³/s
Ayn Sluq	?
Ayn Arus	See below.
Ras al-Ayn	See Table 10.5.

North Central

Few in number.
Water quality apparently good.

Ayn al-Qerd	Very small
Qamishli	Mainly in Turkey

	Natural Flow (m³/s)		After Irrigation Flow (m³/s)		Total Flow (m³/s)
	Spring	Surface	Spring	Surface	
Qamishli	3	2	1 or 2	2	3 or 4
Ayn Arus	6	?	2 est.		2

Source: FAO (1966, 12, 26–27, 195).

lent porosity and permeability and, where either precipitation or infiltration from streams is available, provide good stores of immediately available groundwater for the upper saturated zone. These formations are of particular importance to the east and southeast of Qamishli, where they acquire waters of the Jagh Jagh, the Brebich-Jarrah, and the Roumelie and, in turn, release large amounts into the Radd for subsequent evaporation.

Turkish-Syrian Shares of Khabur Waters

The above description of the Khabur basin provides the basis for an analysis of how water is utilized within the basin, both where it comes from and where it goes.[1] Obviously, the Khabur is an independent system receiving no water from the Euphrates but contributing to the larger stream. Therefore, precipitation is considered to be the sole source of water passing through the system. The geologic structures mentioned previously preclude the addition of underground waters from outside the topographic basin. On the other hand, the subsystems of the Khabur, each within its own smaller drainage area, exchange water both above and below the ground with adjoining sub-basins.

The basic problem facing this analysis was twofold: to assign amounts of precipitation to the Turkish and Syrian segments of the system and to assign final values regarding runoff in the same way, but also to take into account differences in evapotranspiration and use from one place to another.

Table 9.4 presents the first half of this task. Sections of each sub-basin were carefully measured and assigned to either Turkey or Syria. In turn, the precipitation falling on each area was calcu-

---

1. While many sources have been consulted during the analysis and writing of the materials presented here, one above all has provided the necessary background information. This is the *Études des resources en eaux souterraines de la Jezirah Syrienne* prepared by the Food and Agriculture Organizations of the United Nations in cooperation with the Government of Syria (1966). This undertaking covered the full spectrum of subject matter from basic climatology and geology to land use and agricultural economics. While many of the data used within it are of necessity of short time span, the workers exercised the utmost caution and modesty in making their analyses. Much of the material, however, was presented solely in terms of Syrian use of the area. While this was entirely natural and proper, the fact that Turkey may have rival claims to some of the water resources involved was noted but scarcely taken into consideration by the FAO team. It has been necessary, therefore, to rework sections of the report in order to give a more international perspective to the questions involved.

Table 9.4. Coefficients of Precipitation in the Basin and Sub-Basins of the Khabur River

Description of sub-basin	Total Area km^2	Avg Precip per year mm	Total Precip per year 1000s m^3	Area in Turkey km^2	Avg Precip Turkey mm	Total Precip Turkey 1000s m^3	% Water-shed in Turkey	Area in Syria km^2	Avg Precip Syria mm	Total Precip Syria 1000s m^3	% Precip from Turkey	% Precip from Syria
A: Khabur to Ras al-Ayn	3,175	466	1,479,550	3,175	466	1,479,550	100	—	—	—	100	—
B: Djirdjib to confluence with the Khabur	2,775	495	1,371,775	2,540	510*	1,295,400	91.5	235	325*	76,375	94.4	5.6
C: Khabur Basin between R. a-A. and Tel Tamer	1,500	263	394,500	—	—	—	—	1,500	263	394,500	—	100
D: Zergane to Tel Tamer	2,575	470	1,208,052	1,822	525*	956,550	70.8	753	334*	251,502	79.2	20.8
SUBTOTAL: Khabur to T. T.	10,025	455	4,453,877	7,537	xx	3,731,500	75.2	2,488	xx	722,377	83.8	16.2
E: Khabur Basin between T. T. and Hasakah	1,000	282	282,000	—	—	—	—	1,000	282	282,000	—	100
SUBTOTAL: Kh. to Hasakah	11,025	430	4,735,877	7,537	xx	3,731,500	69.4	3,488	xx	1,004,377	78.8	21.2

F: Jagh Jagh to Qamishli	1,025	596	610,900	1,025	596	610,900	100	—	—	—	100	—
G: J. J. between Qamishli and Sfaya	10,800	384	4,190,400	2,160	500*	1,080,000	20.0	8,640	360*	3,110,400	25.8	74.2
H: J. J. between Sfaya and Hasakah	675	311	209,925	—	—	—	—	675	311	209,925	—	100
SUBTOTAL: J. J. Basin to Hasakah	12,500	xx	5,011,225	3,185	xx	1,690,900	25.5	9,315	—	3,320,325	33.7	66.3
I: Khabur Basin between Hasakah and Suwar	7,675	222	1,703,850	—	—	—	—	7,675	222	1,703,850	—	100
TOTAL: The Khabur to Suwar	31,200 100%	366	11,450,952 100%	10,722 34.4%	506	5,422,400 47.4%	34.4	20,478 65.6%	294	6,028,552 52.6%	47.4	52.6

Source: Based on FAO (1966, 66–67, Tables III-5 and III-6, and on Map 1, endpapers.

*Estimate made from FAO materials

lated and weighted according to north-south variations in annual amounts. The last two columns on the right of this table present the calculated amounts of precipitation in each sub-system for each country. Such percentages can then serve as a means of weighing the amount of runoff from each sub-system. (It should be noted that this table has an internal means of balancing its values which may be summed from top to bottom.)

What becomes apparent may at first seem somewhat anomalous. That is, only 34 percent of the basin and 47 percent of the precipitation are found within Turkey. Yet all the discussion to this point implies that Turkey is the predominant supplier of water to the system. This can be explained and verified with reference to two facts. Average precipitation in the pertinent portions of Turkey is 506 mm/yr while that in the Syrian portion is only 294 mm. Second, evapotranspiration is significantly greater in Syria. In large tracts of the latter country included in this analysis, even in the rainiest month of the year, evapotranspiration exceeds precipitation with no resulting surplus to runoff. In those cases, where average precipitation figures were lacking, proportional estimates based on spatial distributions were used (FAO 1966, map I, endpapers).

Table 9.5 provides a detailed analysis of water use in each sub-basin. (It should be noted that sections of the tributaries analyzed separately in Table 9.1 have been aggregated in Table 9.2. Capital letters identify such groupings.) Because of the complexity of the data, sub-basins shown on Map 9.6 have been stylized for clarity on Map 9.7 and laid out schematically on Figure 9.2.

In order to explain the analysis, the following description traces row "a" from left to right. (The following explanation may also be used with Figure 9.2.)

F	This provides the descriptive location of the river segment referred to in Table 9.1.
P	FAO data indicated that this area provided 7 m^3/s to the system.
G-W$_1$	Of these 7 m^3/s, 2.5 infiltrated into groundwater and/or aquifers.
G-W$_2$	At the same time, 2 m^3/s entered the sub-

	basin from the Jagh Jagh between Qamishli and Sfaya. This latter exchange is between sub-systems and must be accounted for separately.
R	Surface flow removed another 2.0 m³/s downstream.
S	Another 1 m³/s of spring flow also moved downstream.
UF	A similar sub-system exchange of underflow in the river alluvia removes 1.5 m³/s into the Jagh Jagh between Qamishli and Sfaya. There is an apparent two-way exchange of underflow and groundwater in this area. The end result is a net loss of 0.5 m³/s from the Jagh Jagh at this point. (See G-W₁ above.)
E_m, E_m-s	In this case no water is lost by evapotranspiration from marshes or semi-marshes although in other sub-systems, such is the case.
E_i	Irrigation removes 2 m³/s from the system through evapotranspiration losses.
Total In/Out	Summing the pluses and minuses balances this row.
$R+E_i+S$	The *natural flow* of this sub-system is equal to that from rivers and springs plus what is lost through human activity. (Sub-system exchanges are accounted for in other subsections.) The amount of water entering the Khabur from this sub-basin is equal to 5 m³/s.
% from Turkey	Since 100 percent of the precipitation—the source of the above flow—has been shown in Table 9.1 to have come from Turkey, 5 m³/s have been assigned to Turkey.

The conclusions reached by this accounting show that 47.7 m³/s of the natural flow of the Khabur and its tributaries should be assigned to Turkey as surface runoff or from aquifers whose

Table 9.5. Turkish-Syrian Shares of Available Water—Allocation of Precipitation in the Khabur Basin, Sub-Basins, and Catchment Area*

Hydro-area	Basin Area Equivalent See: Table 10.1	Precipi-tation P	Ground-water G-W	Surface Flow (Rivers) R	Surface Flow (Springs) S	Under-flow UF	Marshes E_m	Semi-Marshes E_{s-m}	Irrig (1961) E_i	Total in	Total out	Nat. flow $R+E_i$ +S	% from Turkey (Table 10.1)	Nat. flow orig. in Turkey
a	F	7.0	−2.5	−2.0	−1.0	−1.5	—	—	−2.0	+9.0	−9.0	5.0	100	5.0
	The Jagh Jagh													
	to		+2.0			subsystem								
	Qamishli					exchange								
b	G	15.5	−2.0	−2.5	—	+1.5	−2.0	−3.0	−0.5	+17.0	−17.0	3.0	25.8	.8
	J. J. between													
	Qam. and		subsystem				−3.0	−4.0						
	Sfaya		exchange											
c	H	4.5	−2.0	−1.0	—	—	—	−1.5	—	+4.5	−4.5	1.0	—	—
	J. J. between													
	Sfaya and													
	Hasakah													
SUBTOTAL	Jagh Jagh													
	System	27.0	−4.5	−5.5	−1.0	—	−5.0	−8.5	−2.5	+27.0	−27.0	9.0	xx	5.8
										(+30.5)	(−30.5)	(includes subsystem exchange)		
d	A	33.0	+2.5	−5.0	−42.0	—	−1.0	—	−3.0	+51.0	−51.0	50.0	83.8	41.9
	Khabur to													
	Ras al-Ayn													
	"Geol"	13.5	+2.0											

B	Djirdjib to the Khabur													
C	Ras al-Ayn to Tel Tamer													
D	The Zergane to Tel Tamer													
e	E Khabur between Tel Tamer and Hasakah	0.5	—	+3.5 from upstream flow	—	−0.5	—	—	−3.5	+4.0	−4.0	0.0	—	—
—	I Hasakah to Suwar	2.0	—	+1.5	—	—	—	−3.5 includes E_i	—	+3.5	−3.5	−1.5	—	—
SUBTOTAL		49.0	+4.5	—	−42.0	−0.5	−1.0	−3.5	−6.5	+53.5 (+58.5)	−53.5 (−58.5)	48.5 (includes subsystem exchanges)	xx	41.9
BASIN TOTAL BELOW SUWAR		76.0	—	−5.5	−43.0	−0.5	−6.0	−12.0	−9.0	+76.0	−76.0	57.5	xx	47.7

TABLE SUMMARY

Natural Flow:
Flow originating in Turkey = 47.7 = 1.5×10^9 m³/yr.
Flow originating in Syria = 9.8 = $.3 \times 10^9$ m³/yr.
Total flow = 57.5 = 1.8×10^9 m³/yr.

47.7/57.5 = 83% of total flow originates in Turkey.

Source: Based on information in FAO (1966, Chapter 9 and Table K-1).

*Figures in m³/s

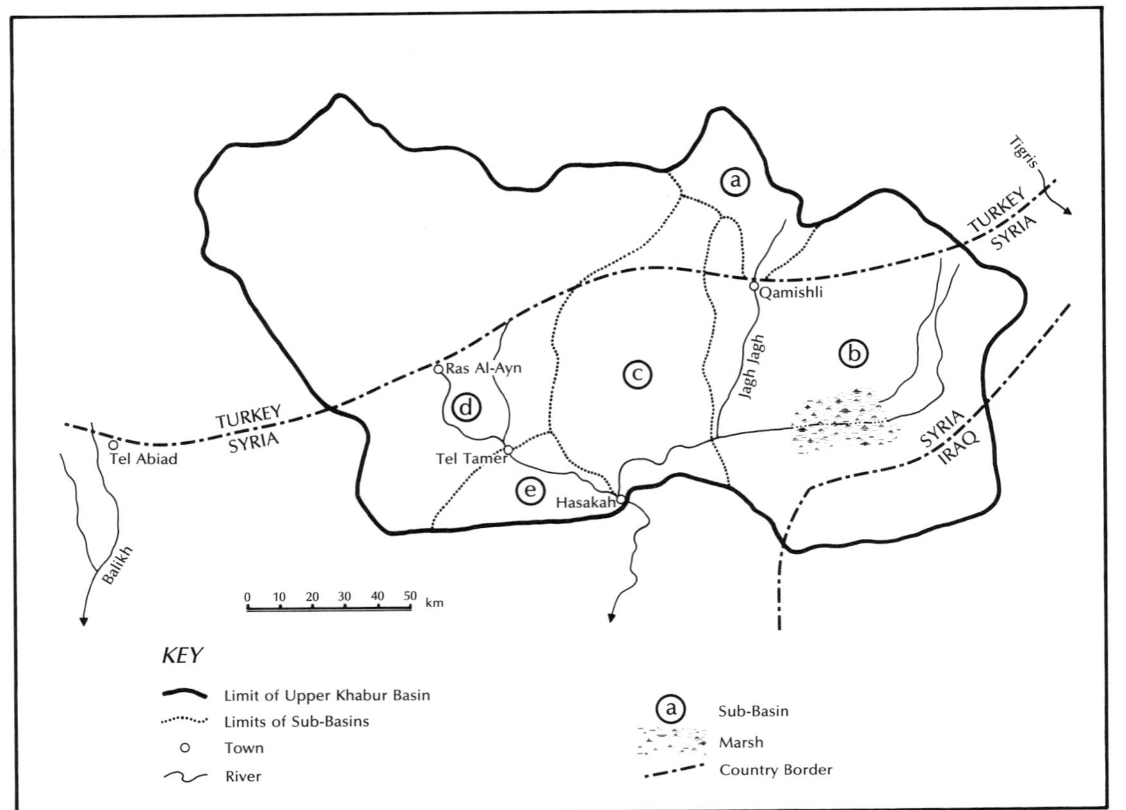

Map 9.6. Sub-drainage basins of the Khabur River. Source: FAO (1966, 191).

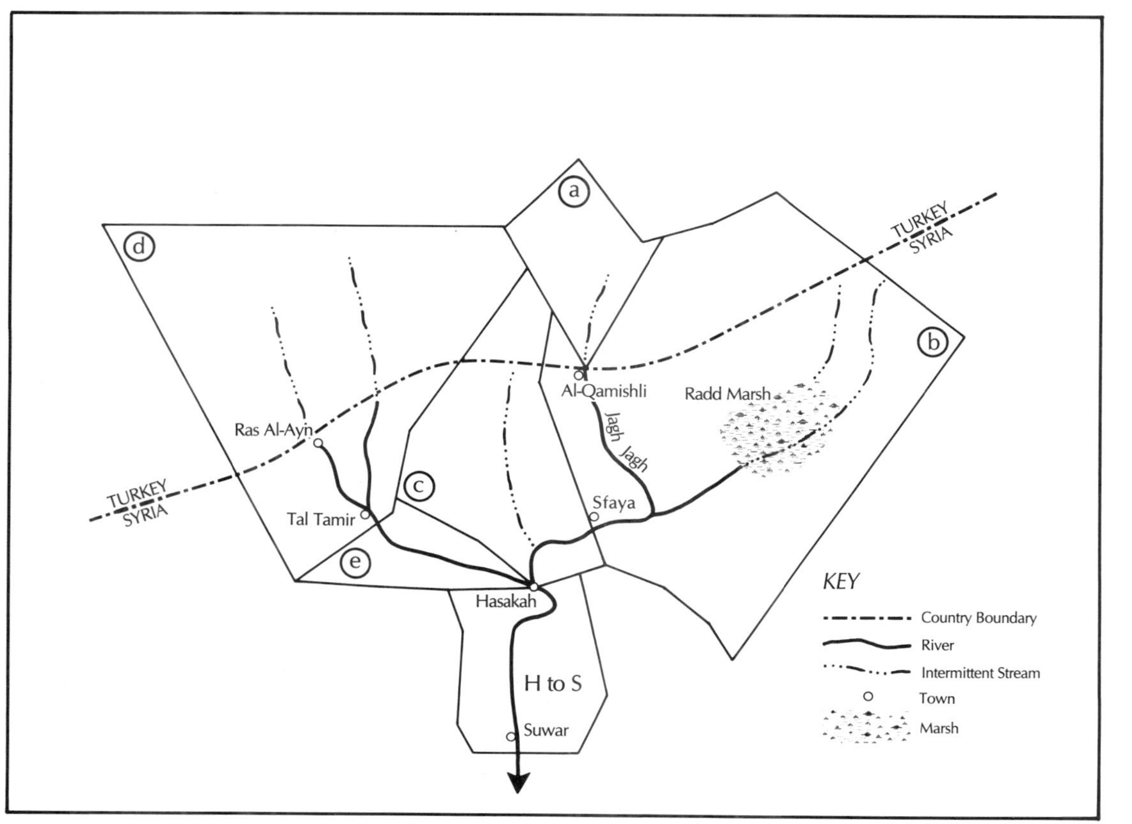

Map 9.7. Hydrologic subdivisions of the Khabur River basin.

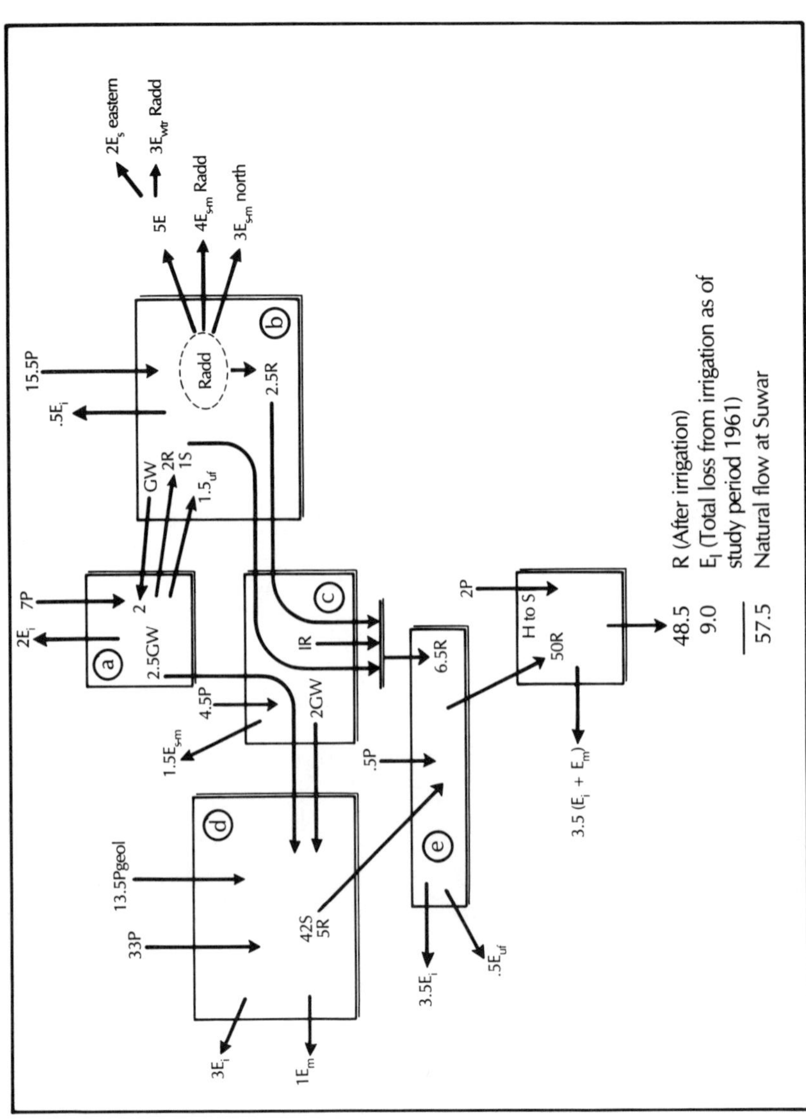

Figure 9.2. Hydrology of the Khabur River basin by subdivisions.

E_i	Evapotranspiration from irrigation
P	Precipitation
P_{geol}	Additional precipitation as groundwater
GW	Groundwater
R	Surface runoff from precipitation
S	Additional discharge from springs
$—_{uf}$	Underflow in stream alluvia
E_m	Evapotranspiration from marshes
E_{s-m} north	Evapotranspiration from semi-marshes (seasonal), northern
E_s eastern	Evapotranspiration during summer from eastern Radd
E_{wtr}	Evapotranspiration during winter from main Radd
H to S	Hasakah to Suwar
Circled letters a–e	Hydrologic subdivisions shown on Maps 10.5 and 10.6.

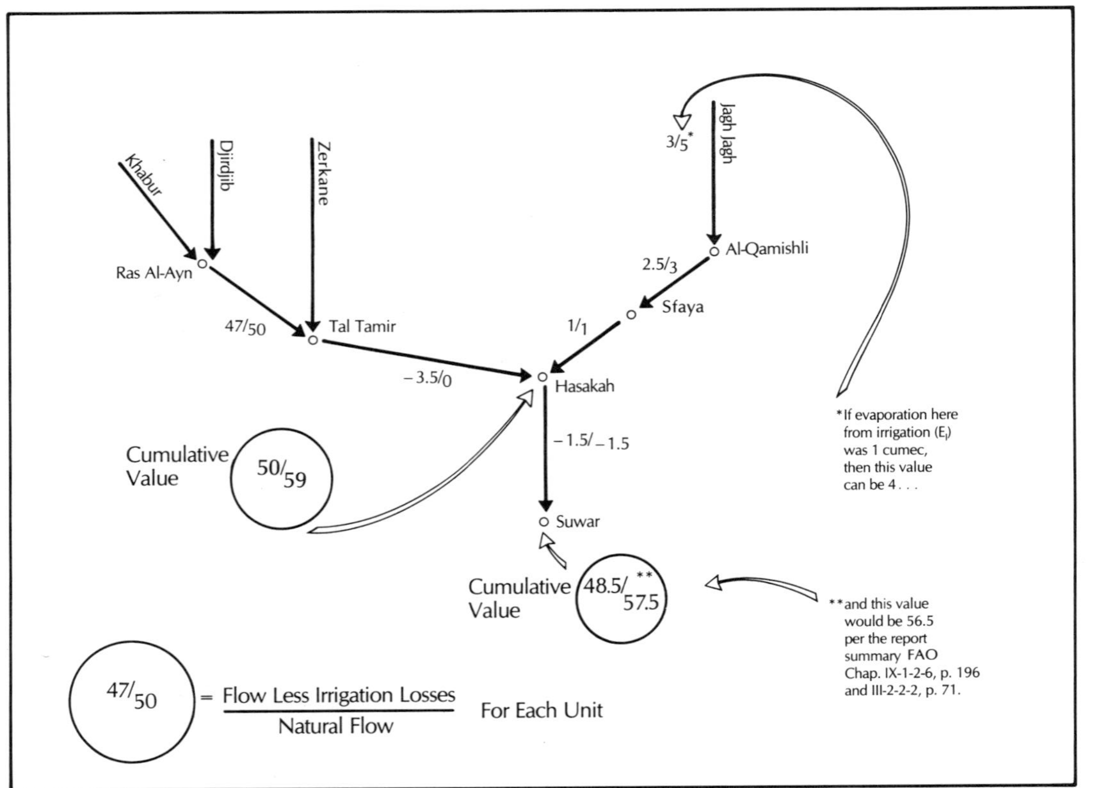

Figure 9.3. Changes per stream unit of actual flow and natural flow of the Khabur River System, 1961.

catchments are in Turkey. Another 9.8 m³/s originate in Syria, making a total of 57.5 m³/s natural flow. *In other words, 83 percent of the total flow of the Khabur originates in Turkey; that is 1,500 Mm³.* Irrigation in Syria removes at least 4.5 m³/s and probably much more of the total 9 m³/s lost. Evapotranspiration from marshes and semi-marshes represents another significant loss which will be considered again in the summary section of the study. (Figure 9.3 further summarizes these remarks.)

If we return to the considerations posed at the beginning of this section, we find a new perspective on the use of water for irrigation in this segment of the Euphrates basin. While a detailed analysis of the Balikh subsystem has not been possible because of lack of data, it may be assumed with considerable certainty that similar amounts of water can be assigned to that portion of Euphrates supply. Indications are that a similar conclusion may be reached regarding the waters of the Sajur to the west. *This means that if roughly 80 percent of the waters named above come in actuality from Turkey, that country's contribution to the total Euphrates system—as demonstrated in Table 6.1—is 29,040 Mm³/yr out of an average of 29,450 Mm³/yr or 98.6 percent!*

This conclusion might be of little importance if it were not for Turkey's plans to establish large-scale pumping of the aquifers to the north of the Syrian border. This may be offset by the return flow from Turkish fields which promises to be great. However, such a return flow, as has been mentioned previously, might well bring new problems of pollution to downstream areas. This will also be considered in the next chapter.

10

Static and Dynamic Views of the Euphrates River System

The analysis presented in the preceding chapters allows a comprehensive view of the Euphrates system as far as the Syrian-Iraq border. The Tigris River has been excluded from this study because it is scheduled for development at a later date. This omission does not preclude the necessity of taking the Tigris portions of GAP into consideration (GAP 1980), but to do so would double the length of this study. Moreover, far fewer data are readily available concerning that stream and its more complex regime vis-a-vis the flow of the left bank tributaries coming from the Zagros Mountains. In the same manner, for all the above reasons and also because the current political situation there would again slow the analysis of conditions in that country, this study touches upon the situation in Iraq but does not consider it in great detail.

Constraints on Dynamic Modeling

A caveat must be made regarding this analysis. When, and if, the development plans of all three countries are in place and functioning, the fair and efficient management of the river will be an exceedingly complex operation. Each of the three riparian users intends to utilize the river for both hydroelectric production and irrigation. Even within the boundaries of a single user, balancing these needs is no small task. For example, Raif Ozenci, the local deputy manager of DSI at the Ataturk Dam site, points out (*Turkish Daily News*, 1984b) that while the Ataturk Dam is designed to produce 8.9 billion kWh annually, electric production will be reduced to 8.1 billion kWh/yr when the proposed irrigation projects come on line. Furthermore, a significant quantity of the power produced will be used locally for pumping water to project fields. These demands will have to be balanced against upstream hydroelectric production at Karakaya and Keban and similar production plus removals for irrigation downstream at Birecik and Karkamis.

Operating in counterpoint to all such variations will be the changes in natural stream flow tied to climatic variations. Moreover, beyond the initial dead-water storage filling of reservoirs, there will be annual fluctuations induced by human needs. The question remains: should water be reserved for uninterrupted hydroelectric production or released both for irrigation purposes and to insure reservoir capacity in the event of unexpected flood conditions?

Table 10.1 illustrates the intricacies of such questions in terms of a situation which has already taken place. Average monthly flow of the Euphrates at Keban is shown in the left-hand column. This varies from a maximum in April of 5,127 Mm^3 to a minimum of 562 Mm^3 in September. Making the impractical and politically unrealistic assumption that *all* the flow of the stream will be held back until the reservoir is filled, considerable variation in the length of time necessary to reach total reservoir capacity occurs depending upon the month of the year in which filling begins. If the gates are closed in March or April, capacity will be reached in May or June of the following year. If the gates were to be closed in June, capacity would not be reached until early in April *23 months later.*

This variation is determined by whether or not spring floods can be retained at optimum times. Obviously, all the water in the river cannot be withheld from downstream users in these cases, but the two confrontations between Iraq and Syria over shortages in the flow of the river which have already been mentioned indicate the delicacy of such timing. In a similar vein, year-to-year

Table 10.1. Keban Reservoir—Recharge Rates
(top capacity = 30,500 Mm³)

Month	Ave. Flow (Thousands of Mm³)	Jan	Feb	Mar	Apr	May	Jun	Jul	Aug	Sep	Oct	Nov	Dec
Jan	.774	.77	—	—	—	—	—	—	—	—	—	—	—
Feb	.890	1.66	.89	—	—	—	—	—	—	—	—	—	—
Mar	1.900	3.56	2.79	1.90	—	—	—	—	—	—	—	—	—
Apr	5.127	8.69	7.92	7.03	5.13	—	—	—	—	—	—	—	—
May	4.802	13.49	12.72	11.83	9.93	4.80	—	—	—	—	—	—	—
Jun	2.053	15.55	14.77	13.88	11.98	6.86	2.05	—	—	—	—	—	—
Jul	.970	16.52	15.74	14.85	12.95	7.83	3.02	.97	—	—	—	—	—
Aug	.659	17.18	16.40	15.51	13.61	8.48	3.68	1.63	.66	—	—	—	—
Sep	.562	17.74	16.96	16.07	14.17	9.05	4.24	2.19	1.22	.56	—	—	—
Oct	.667	18.40	17.63	16.74	14.84	9.71	4.91	2.86	1.89	1.23	.67	—	—
Nov	.783	19.19	18.41	17.52	15.62	10.50	5.69	3.64	2.67	2.01	1.45	.78	—
Dec	.812	20.00	19.23	18.33	16.44	11.31	6.51	4.45	3.48	2.82	2.26	1.60	.81
Jan	.774	20.77	20.00	19.11	17.21	12.08	7.28	5.23	4.26	3.60	3.04	2.37	1.59
Feb	.890	21.66	20.89	20.00	18.10	12.97	8.17	6.12	5.15	4.49	3.93	3.26	2.48
Mar	1.900	23.56	22.79	21.90	20.00	14.87	10.07	8.02	7.05	6.39	5.83	5.16	4.38

Apr	5.127	28.69	27.92	27.03	25.13	20.00	15.20	13.14	12.17	11.52	10.95	10.29	9.50
May	4.802	33.49	32.72	31.83	29.93	24.80	20.00	17.95	16.99	16.32	15.76	15.09	14.31
Jun	2.053	—	—	—	31.98	26.85	22.05	20.00	19.03	18.37	17.81	17.14	16.36
Jul	.970	—	—	—	—	27.82	23.02	20.97	20.00	19.34	18.78	18.11	17.33
Aug	.659	—	—	—	—	28.48	23.68	21.63	20.66	20.00	19.44	18.77	17.99
Sep	.562	—	—	—	—	29.05	24.24	22.19	21.22	20.56	20.00	19.33	18.55
Oct	.667	—	—	—	—	29.71	24.91	22.86	21.89	21.23	20.67	20.00	19.22
Nov	.783	—	—	—	—	30.50	25.69	23.64	22.67	22.01	21.45	20.78	20.00
Dec	.812	—	—	—	—	—	26.51	24.45	23.48	22.82	22.26	21.59	20.81
Jan	.774	—	—	—	—	—	27.28	25.23	24.26	23.60	23.04	22.34	21.59
Feb	.890	—	—	—	—	—	28.17	26.12	25.15	24.49	23.93	23.26	22.48
Mar	1.900	—	—	—	—	—	30.07	28.02	27.05	26.39	25.83	25.16	24.38
Apr	5.127	—	—	—	—	—	35.20	33.14	32.17	31.51	30.95	30.29	29.50
May	4.802	—	—	—	—	—	—	—	—	—	—	35.09	34.30

Source: al-Hadithi (1978).

variations in flow resulting from climatic changes can create difficult situations as shown by Figure 5.2.[1]

Table 10.2 represents a further complication in river management resulting from monthly variation in evaporation rates from reservoir surfaces. Such evaporation is in turn a function of the size of the surface involved. Since reservoirs will be changing volume and surface area depending upon natural conditions and human demands, evaporation losses vary considerably. For example, given a maximum volume of 30,500 Mm^3 in the Keban reservoir, its surface area would be 675 sq km with evaporation losses of approximately 1,000 Mm^3/yr. If the minimum operating level were maintained, volume would be 9,500 Mm^3 with a surface area of 260 sq km and an annual loss of 390 Mm^3/yr. It can safely be assumed that volume and surface area will vary throughout each year and that, therefore, evaporation losses will follow such changes as well as reflecting annual conditions of temperature, wind turbulence, humidity, cloud cover, etc.

The above considerations dictate the ultimate necessity of a dynamic model of the river, but also preclude such an attempt given the limited resources and time allowed for this study. One suggestion is that LANDSAT or other imagery made available on an ongoing basis could provide surface areas of both large and small reservoirs. These, in turn, could be translated into volumes and flow and evaporation rates. Once such an analytical system were in place, river management and surveillance would become considerably simpler.

A Static Model of the Euphrates River and Its Uses

Given the above considerations, what remains possible is a static model or picture of the river using average data which have been

---

1. In the same interview cited above Ozenci also is cited as saying that ". . . it was hoped the year-long process of filling the lake [Ataturk Reservoir] would start in late 1988."(*TDN*, 2 Oct, 1986, 3-1821). It is to be hoped that more than one year will be used in reality for this task.

discussed in previous chapters with approximations of demands for several time periods.

A further clarification becomes necessary regarding the variation shown in the approach to the Turkish and Syrian sections of the Euphrates. Successful development must take account of a riverine-irrigation and economic-political-social conditions. Technology can mean dams and reservoirs, delivery systems, and the application of water to soil. Economic, political, and social conditions concern sufficient funding, agreements between users and between users and suppliers, and the appropriateness of the fit between that which is technologically available and the society which will use or reject it.

In the Turkish case, the ability of the Turks to provide dams and reservoirs has been clearly shown by the time of this writing. Although technological difficulties remain, e.g., the completion of the Urfa tunnels, there seems little reason to suppose that given time and money they will not be overcome. As just indicated, the creation of delivery systems remains to be proven, but again, this seems possible, all else being equal. By the same token, the application of the water to the soil promises to become reality.

As will be described in the final chapter of this study, irrigated agriculture is already underway from underground sources in the Harran Plain and near Ceylanpinar. Moreover, the Class I and II soils upon which the bulk of Turkish GAP development depends are suitable for irrigation. Therefore, from a technological point of view, the GAP sub-projects discussed in this study and further to be examined in this chapter should all be realizable—if time, money, and politics and society allow.

On the other hand, the development of irrigation in Syria appears to hinge upon the quality of the soils to be irrigated, as well as on problems associated with delivery systems dependent upon those soils. In fact, few people displaced by reservoir flooding have been resettled, and, as will be seen, glowing projections of large tracts soon to be irrigated have steadily been diminishing in size. Moreover, there is considerable confusion as to just how much land can be irrigated. It seems that, because of gypsiferous soils, what is actually happening is that private irrigation of Valley soils is slowly giving way to centrally controlled irrigation of

Table 10.2. Keban Reservoir Average Evaporation and Average Inflow

	Jan	Feb	Mar	Apr	May	Jun	Jul	Aug	Sep	Oct	Nov	Dec	Annual
Inflow Ave. (in m³)	774	890	1900	5127	4802	2053	970	659	562	667	783	812	19,999
Evap. Ave. (mm)	15.5	8.1	56.5	96.6	159.8	215.6	290.6	284.8	174.4	113.3	48.8	21.1	1484.6

Evaporation by Area and by Month
(In Mcm)

Elev. (m)	Area (km²)	Vol.	Jan	Feb	Mar	Apr	May	Jun	Jul	Aug	Sep	Oct	Nov	Dec	Annual
FULL SUPPLY															
845	675	30,500	10.4	5.5	38.1	65.2	107.9	145.5	196.2	192.2	117.7	76.5	32.9	14.2	1002.4
840	620	27,000	9.6	5.0	35.0	59.9	99.1	133.7	180.1	176.5	108.1	70.3	30.2	13.1	920.6
NORMAL LEVEL															
835	570	24,200	8.8	4.6	32.2	55.1	91.1	122.9	165.6	162.3	99.4	64.6	27.8	12.0	846.4
830	525	21,700	8.1	4.3	29.7	50.7	83.9	113.2	152.6	149.5	91.6	59.5	25.6	11.1	
825	480	19,200	7.4	3.9	27.1	46.4	76.7	103.5	139.5	136.7	83.7	54.4	23.4	10.1	
818	430	16,000	6.7	3.5	24.3	41.6	68.7	92.7	125.0	122.4	75.0	48.7	21.0	9.1	
815	385	14,600	6.0	3.1	21.8	37.2	61.5	83.0	111.9	109.7	67.1	43.6	18.8	8.1	
805	300	11,000	4.7	2.4	17.0	29.0	47.9	64.7	87.2	85.4	52.3	34.0	14.6	6.3	

MIN. OPERATING LEVEL

800	260	9500	4.0	2.1	15.4	25.1	41.5	56.0	75.5	74.0	45.3	29.5	12.7	5.5	386.7
794	225	8000	3.5	1.8	12.7	21.7	36.0	48.5	65.4	64.1	39.2	25.5	11.0	4.8	
784	180	6000	2.8	1.5	10.2	17.4	28.8	38.8	52.3	51.3	31.4	20.4	8.8	3.8	
777	160	5000	2.5	1.3	9.0	15.5	25.6	34.5	46.5	45.6	27.9	18.1	7.8	3.4	
772	140	4000	2.2	1.1	7.9	13.5	22.4	30.2	40.7	39.9	24.4	15.9	6.8	3.0	
760	107	2800	1.7	0.9	6.1	10.3	17.1	23.1	31.1	30.5	18.7	12.1	5.2	2.3	
753	90	2000	1.4	0.7	5.1	8.7	14.4	19.4	26.2	25.6	15.7	10.2	4.4	1.9	
746	75	1500	1.2	0.6	4.3	7.3	12.0	16.2	21.9	21.4	13.1	8.5	2.6	1.6	110.7
738	55	1000	0.9	0.5	3.1	5.3	8.8	11.9	16.0	15.7	9.6	6.2	2.4	1.2	
734	45	800	0.7	0.4	2.5	4.4	7.2	9.7	13.1	12.8	7.9	5.0	2.2	1.0	
720	20	300	0.3	0.2	1.1	1.9	3.2	4.3	5.8	5.7	3.5	2.3	1.0	0.4	
700	0	0	0.0	0.0	0.0	0.0	0.0	0.0	0.0	0.0	0.0	0.0	0.0	0.0	

Source: Based on al-Hadithi (1978, Tables 8, 9, and 13) but based on a 365-day year.

the same plots with little new upland soil being brought into production. (The one exception to this may be the projects along the upper Khabur about which more will be said.)

This is not to say that the Syrian effort is without economic, social, and political problems, but rather that, in the area of technical feasibility, significant differences appear when comparing the two nations' projects.

Thus the following attempt to quantify the amount of land and water ultimately to be used by the Turks and the Syrians makes two assumptions based upon the technological aspects of the situation: first, that the areas designated by the Turks for irrigation can and will receive water at some future date and, second, that initial Syrian proposals have been and will be seriously reduced in size for the reasons stated above.

The compilation shown in Table 10.3 begins with the original amounts cited by both the Turks and the Syrians. In the latter case, however, Chapter 11 considers what may ultimately be expected of the Syrian development program, something considerably less than their initial proposals. Figure 10.1, showing the impact of water removals from the river over time, presents the original Turkish estimates throughout but provides for consideration several levels of removal from the Syrian section of the river.

A word on the values used in the computations: Without detailed on-site measurements of a number of variables, the values used in computing evaporation, evapotranspiration, irrigation water needs, conveyance and farm efficiency, and return flow and water quality must be based on available data and intelligent estimates of conditions. Evapotranspiration and irrigation water needs have been discussed at length in previous chapters of this book. Evaporation from reservoir surfaces has been computed using values provided by al-Hadithi (1979); these fall within the expected range.

Return flow values shown in Appendix A, while falling within a relatively narrow range, have presented some difficulty in making a final choice for this analysis. A round figure of 35 percent has been chosen. This is perhaps generous, and Table 10.4 shows the consequences of choosing a more conservative 30 percent and 25 percent for selected cases. On the one hand, this relatively high

Table 10.3. Officially Anticipated or Enacted Dams, Reservoirs, and Irrigation on the Euphrates River, 1986–2000+

Part I: Headwaters to the Syrian Border

In Operation Ca 1986–1990

Project Name (River Name)	Status (Vol. in Mm³)	Res. Area km² (Evap. Rate m/yr)	Water depletion per yr in Mm³	Irrig. Area ha (depletion rate in m³/ha/yr)	Ave. Water depletion per yr in Mm³	Return Flow in Mm³	Projected Evaporation From Reservoir and Loss From Fields Mm³/yr				Data Source
							1986–90	1990–95	1995–2000	2000+	
Tercan (Tuzla)	op (178)	8.85 (1.2) est.	10.62	32,000 (6642) est.**	212.544	114.5*	223.16				DSI (1984a)
Kalecik (Kalecik)	op (12.5)	1.16 (1.2) est.	1.39	1300 (6642) est.	9.635	4.7	11.00				(DSI 1984a)
Cip (Cip)	op (7.0)	1.10 (1.2) est.	1.32	800 (6642) est.	5.31	2.9	6.63				DSI (1984a)
Gayt (Gayt)	UC (23)	2.92 (1.2) est.	3.50	3200 (6642) est.	21.21	11.5			24.75		DSI (1984a)
Mercan-HEPP	FDC			—	—	—					DSI (1984a)
Girlevik-HEPP	op			—	—	—					DSI (1984a)
Hazar I-HEPP	op			—	—	—					DSI (1984a)

*RF = 3577/ha.
**See Table 8.4 estimated values based on Siverek value.

continued

In Operation Ca 1986–1990

Project Name (River Name)	Status (Vol. in Mm³)	Res. Area km² (Evap. Rate m/yr)	Water depletion per yr in Mm³	Irrig. Area ha (depletion rate in m³/ha/yr)	Ave. Water depletion per yr in Mm³	Return Flow in Mm³	Projected Evaporation From Reservoir and Loss From Fields Mm³/yr				Data Source
							1986–90	1990–95	1995–2000	2000+	
Hazar II-HEPP	op		—	—	—						DSI (1984a)
Sekerova (Badisan)	UFD (90.2)	3.81 (1.2) est.	4.57	15,938 (6642) est.	105.86	57.0*			110.4		DSI (1984a)
Patnos (Gevi)	UFD (33.4)	4.65 (1.2) est.	5.58	4993 (6642) est.	33.16	17.6			38.7		DSI (1984a)
Ozluce-HEPP (Peri Suyu)	op (1075)	6.2 (1.2) est.	31.44	—	—	—	31.4				DSI (1984a)
Keban-HEPP (Euph.)	op (30,600)	675.0 (1.46) AL-H.	985.5	—	—	—	985.5				DSI (1984a); GAP (1980)
Mursal (Hikme)	UFD (17.6)	62 (1.46) est.	2.37	1665 (8081) est.	13.45	7.24**			15.82		DSI (1984a)
Medik (Tohma)	op (22.0)	1.62 (1.46) est.	2.37	15,800 (8081) est.	127.7	68.7		130.1			DSI (1984a)
Yazihan (Tohma)	UC ?	?		9500 (8081) est.	76.77	41.3		76.8[t]			(TSI-1980) ?
Sultan Suyu	UFD	2.6	3.80	14,963	120.9	65.1		124.7			DSI (1984a)

*RF = 3577/ha.
**RF = 4351/ha.

continued

Table 10.3. (continued)

In Operation Ca 1986–1990

Project Name (River Name)	Status (Vol. in Mm^3)	Res. Area km^2 (Evap. Rate m/yr)	Water depletion per yr in Mm^3	Irrig. Area ha (depletion rate in m^3/ha/yr)	Ave. Water depletion per yr in Mm^3	Return Flow in Mm^3	Projected Evaporation From Reservoir and Loss From Fields Mm^3/yr				Data Source
							1986–90	1990–95	1995–2000	2000+	
(Sultan Suyu)	(53.3)	(1.46) est.		(8081) est.							
Kermek-HEPP (Tohma?)	op			—	—	—					DSI (1984a)
Tohma-HEPP (Tohma)	op			—	—	—					DSI (1984a)
Karakaya-HEPP (Euph.)	Filling June '86 (9580)	298.0 (1.46) AL-H.	435.1	—	—	—	435.1				DSI (1984a); *MEED*(?)
Cat (Abdulharap)	op (240)	14.3 (1.6) est.	22.9	22,091 (8954)	197.8	106.5*	220.7				DSI (1984a); TDN(?)
Adiyaman/Kahta											
Adiyaman (Kahta)	UC? (617)	?		80,000 (8954)	716.3	385.7		716.3			GAP (1980)
Kahta (Kahta)	UC? (1887)	?		80,000 (8954)	716.3	385.7		716.3			GAP (1980)
Cankara (Goksu)				38,420 (8954)	344.0	185.2		344.0			GAP (1980)

*RF = 4821/ha.

continued

Table 10.3. (continued)

In Operation Ca 1986–1990

Project Name (River Name)	Status (Vol. in Mm³)	Res. Area km² (Evap. Rate m/yr)	Water depletion per yr in Mm³	Irrig. Area ha (depletion rate in m³/ha/yr)	Ave. Water depletion per yr in Mm³	Return Flow in Mm³	Projected Evaporation From Reservoir and Loss From Fields Mm³/yr				Data Source
							1986–90	1990–95	1995–2000	2000+	
Lower Euphrates Project											
Ataturk (Euph.)	UC (48,700)	817.0 (1.80) Al-H.	1470.6	see below	—	—	1470.6				GAP (1980)
Urfa/Harran	UC	—	—	157,000 (10,754)	1688.4	909.2*	1688.4				GAP (1980, V-8-14 for all Lower Euphrates Project)
Tektek Plateau	UFD	—	—	20,000 (10,754)	215.1	188.2			215.1		
Lower Mardin-Ceylanpinar	UC	—	—	140,000 (11,489)	1608.5	866.0*			1608.5		
Derik-Mardin	UC			192,100 (11,229)	2157.1	1161.4	2157.1				
Derik	(345)	?									
Mardin	(335)	?									
Nusaybin-Cizre	UFD	?		47,000 (11,229)	527.8	284.2			527.8[t]		

*RF = 5,791/ha.   **RF = 6,186/ha.   ***RF = 6,046/ha.

continued

In Operation Ca 1986–1990

Project Name (River Name)	Status (Vol. in Mm³)	Res. Area km² (Evap. Rate m/yr)	Water depletion per yr in Mm³	Irrig. Area ha (depletion rate in m³/ha/yr)	Ave. Water depletion per yr in Mm³	Return Flow in Mm³	Projected Evaporation From Reservoir and Loss From Fields Mm³/yr				Data Source
							1986–90	1990–95	1995–2000	2000+	
Siverek-Hilvan* (16 small dams and reservoirs)	UFD (407.3)	45.52 (1.6) est.	68.03	164,300 (8954)	1471.1	792.1				1539.1	
Hacihidir	op (67.6)	4.40 (1.6) est.	7.04	3400 (8954)	30.4	16.4	37.4				TDN; GAP (1980)
Dumluca	op (22.06)	2.23 (1.6) est.	3.57	2400 (8954)	21.5	11.6	25.1				TDN (8/30/84)
Bozova (pumped from Ataturk Res.)	—	—	—	55,300 (10,754)	594.7	320.2				594.7	
Suruc-Baziki (Yaylak)	UFD (pumped in large part from Ataturk Res.)										
Baziki	—	—	—	44,900 (10,754)	482.9	260.0				482.9	
Suruc Tozluca	UFD (12.35)	?		101,600 (10,754)	1092.6	588.3				1092.6	
Aylan	UFD (6.95)	?									
Tasbasan	UFD (7.68)	?									

*Of this, 147,000 ha drains to Euphrates; 17,300 ha to Khabur.

continued

Table 10.3. (continued)

In Operation Ca 1986–1990

Project Name (River Name)	Status (Vol. in Mm³)	Res. Area km² (Evap. Rate m/yr)	Water depletion per yr in Mm³	Irrig. Area ha (depletion rate in m³/ha/yr)	Ave. Water depletion per yr in Mm³	Return Flow in Mm³	Projected Evaporation From Reservoir and Loss From Fields Mm³/yr				
							1986–90	1990–95	1995–2000	2000+	Data Source
Euphrates Border Project	FDC										
Birecik-HEPP	FDC (1220)	56.25 (2.0) est.	112.5			See Araban and Gaziantep Projects				112.5	
Karkamis-HEPP	FDC (157)	28.4 (2.0) est.	56.8	—	—	—				56.8	
Araban (Karasu)	FDC —	? (pumped from Birecik Res.)		1610 (10,562)	17.0	9.16	17.0				
				21,738 (10,562)	229.6	123.6	229.6				

continued

Table 10.3. (continued)

In Operation Ca 1986–1990

Project Name (River Name)	Status (Vol. in Mm³)	Res. Area km² (Evap. Rate m/yr)	Water depletion per yr in Mm³	Irrig. Area ha (depletion rate in m³/ha/yr)	Ave. Water depletion per yr in Mm³	Return Flow in Mm³	Projected Evaporation From Reservoir and Loss From Fields Mm³/yr				
							1986–90	1990–95	1995–2000	2000+	Data Source
Gaziantep	(pumped from Birecik Res.)			51,789 est. (10,562)	547.0	294.6				547.0	
Hancagiz (Nizip)	FDC (31.72)	7.5 (2.0)	15	10,736 (10,562)	113.4	66.1			128.4		
Kayacik (Tuzel Suyu that flows into the Sajir)	FDC	?		15,700 est. (10,562)	165.8	89.3			165.8		
Kemlim (Balik which becomes the Queiq/Kweik in Syria but is outside the Euphrates drainage)	FDC (31.72)	?		10,736 est. (10,562)	113.4				113.4[t,x]		

*Does not include losses from pumping of groundwater and aquifers in Lower Euphrates Project.
'Partial value.
ˣNot included in totals.

continued

Table 10.3. (continued)

TOTALS FOR TURKEY

(Does not include values for reservoir areas and irrigation projects the sizes of which are not available.)

Total ha Irrigated Potential	Return Flow Mm^3	Total Estimated Water Depletion* Mm^3/yr			
		1986–90	1990–95	1995–2000	2000+
1,350,243	7434	1976.0	7989.9	2516.32	4425.6
	(post 2000)		9965.9	12,482.2	16907.8

*Does not include losses from pumping of groundwater and aquifers in Lower Euphrates Project.
'Partial value.
ˣNot included in totals.
  op = operational.
  UC = under construction.
  UFD = under final design.
  FDC = final design completed.

continued

Table 10.3. (continued)

Part II: Turkish Border to the Iraqi Border

In Operation Ca 1986–1990

Project Name (River Name)	Status (Vol. in Mm³)	Res. Area km² (Evap. Rate m/yr)	Water depletion per yr in Mm³	Irrig. Area ha (depletion rate in m³/ha/yr)	Ave. Water depletion per yr in Mm³	Return Flow in Mm³	Projected Evaporation From Reservoir and Loss From Fields Mm³/yr				Data Source
							1986–90	1990–95	1995–2000	2000+	
? (Sajur)	? (2)	?		?		?					USAID (1980)
Tishreen Dam (Euph.)	planning (1300)	70 est. (2.25) est.	157.5	?		?		157.5ᵗ			MEED (8/9/80)
Aleppo Diversion	op	—	80.2	—	—	—	80.2				
Lake Assad (Euph.) (Tabqa Dam) *	op (11,700)	628 (2.5)	1570.0	—	—	—	1570.0				al-Hadithi (1978, Table 15)
Maskanah- Aleppo #6 (Euph.)	UC	—	—	150,000 (12,545) est.	1882	1013			1882		
Rasafah #4				abandoned							
Balikh #1 (Balikh)	UC	—	—	185,000 (10,754) est.	1989.5	1071			1989.5		

*The Baath Dam and Reservoir (2.7 km²/evap. 6.75 Mm³/yr) are not included—see Table 9.4.

continued

Table 10.3. (continued)

In Operation Ca 1986–1990

Project Name (River Name)	Status (Vol. in Mm³)	Res. Area km² (Evap. Rate m/yr)	Water depletion per yr in Mm³	Irrig. Area ha (depletion rate in m³/ha/yr)	Ave. Water depletion per yr in Mm³	Return Flow in Mm³	Projected Evaporation From Reservoir and Loss From Fields Mm³/yr				
							1986–90	1990–95	1995–2000	2000+	Data Source
Lower Valley #2 (Euph.)	UC	—	—	165,000 (16,835)	2777.8	1496			2777.8		
Baath Dam (Euph.)	OP ?	? (2.3) est.	?								
Upper Khabur (Khabur & tribs.)	UC (irrig. from Raas Al-Ayn)	—	—	42,000 (11,489)	483	260	483				(see Table I-1)
W. Al-Hasakah	UC (91)	.102 (2.3) est.	23.46	49,450 combined	1203.4	648		1203.4			
E. Al-Hasakah	UC (232)	.310 (2.3) est.	71.30	(12,545)							(see Table I-1)
Al-Khabur	UC (665)	.958 (2.3) est.	220.34	46,450 (12,545)	combined	combined			combined		
Lower Khabur* #3 (Khabur/Euph. = from Lake Assad)	FDC	—	—	75,000 (16,835)	1262.6	679				1262.6	(see Table I-1)
Mayadin Plain* #5 (Euph.)	FDC	—	—	40,000 (16,835) est.	673.4	363				673.4	(see Table I-1)

*Reported as of 1982 to be "abandoned or suspended" (Metral 1987, 119) See Chapter 12 for a further discussion.

continued

Table 10.3. (continued)

TOTALS FOR SYRIA

Total ha Irrigated Potential	Ave. Water Depletion per yr in Mm^3	Return Flow Mm^3	1986–90	1990–95	1995–2000	2000+
752,900** (Planned)		5530 (Planned)	2133.2	1360.9	8585.3	
				3494.1		
					12,079.4	
Cumulative Totals for Turkey			1976.0	9965.9	12,482.2	16,907.8
TOTAL DEPLETIONS TO IRAQI BORDER			4109.2	13,460.0	24,561.6	28,987.2

**See chapter 12 for alternate figures.
'Partial value.
  op = operational.
  UC = under construction.
  UFD = under final design.
  FDC = final design completed.

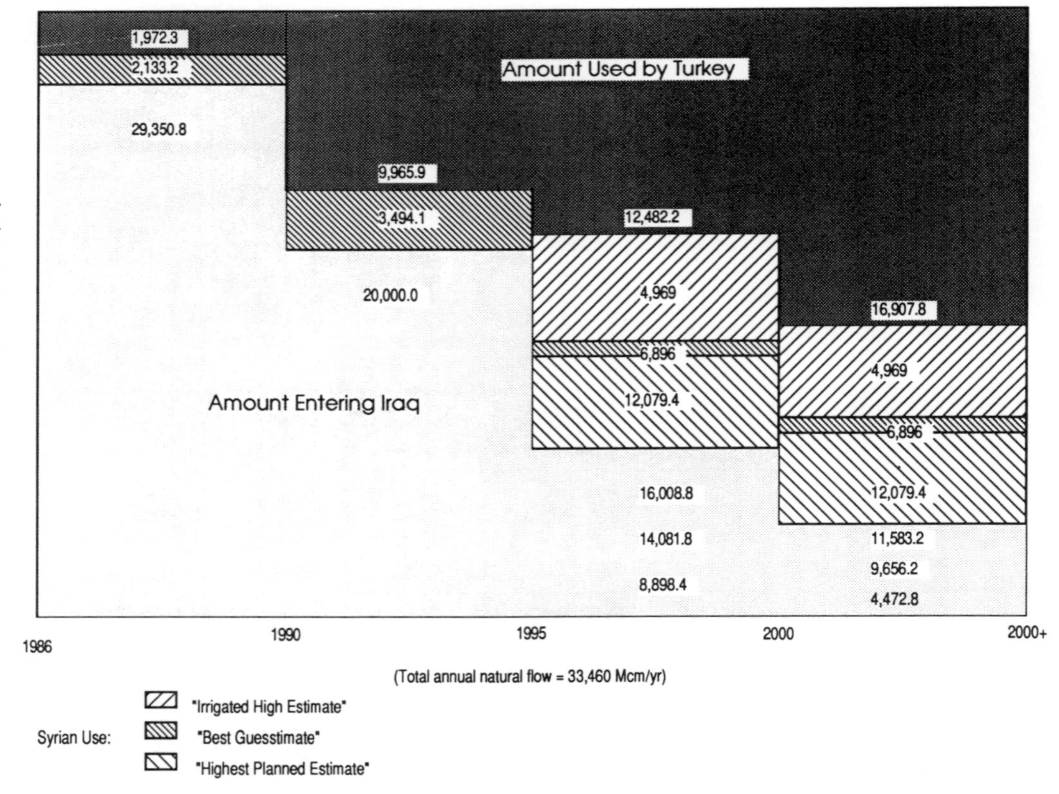

Figure 10.1. Water subtractions Euphrates River: Upstream users optimum scenarios, 1986–2000+ (amounts in Mm^3/yr).

Table 10.4. Variations in Estimated Water Use, Loss, and Depletion as a Function of Values Chosen

Assumed Need	Amount Withdrawn	RF	=	Return Flow	Nonproductive Loss	Comments
10	(2.5)	(0.35)	=	8.75	6.25	selected for
	depletion		=	16.25		this study
10	(2.5)	(0.30)	=	7.50	7.50	
	depletion		=	17.50		
10	(2.5)	(0.25)	=	6.25	8.75	most
	depletion		=	18.75		pessimistic
10	(2.0)	(0.35)	=	7.00	3.00	most
	depletion		=	13.00		optimistic
10	(2.0)	(0.30)	=	6.00	4.00	
	depletion		=	14.00		
10	(2.0)	(0.25)	=	5.00	5.00	
	depletion		=	15.00		

	Selected Value	Most Pessimistic	Most Optimistic
Nonproductive Loss	6.25	8.75	3.00
% Change	0.0	+40%	−52%
Return Flow	8.75	6.25	7.00
% Change	0.0	−28%	−20%

return flow (RF) value gives the benefit of the doubt to upstream users that much of what they remove will find its way to downstream riparians. On the other hand, considering the very large volumes of water involved upstream, RF may present serious problems of flooding, water-logging and/or pollution to downstream users. These are matters that should be resolved first by on-the-spot experts and then through negotiations regarding removal and use rates.

Conveyance and on-farm efficiency and their corollary, water loss, are discussed in Appendix A. For this summary, water lost to the system has been computed in the following manner. Irrigation water needs (the amount of water needed for optimum crop production less the amounts of water provided naturally by effective rainfall and soil moisture recovery), based on Thornthwaite's method of computation, have been used as the base value. (As mentioned earlier, while the Blaney-Criddle method gives a closer *and usually higher* value, data constraints limited this analysis to the use of Thornthwaite. The values given herein may be considered as minimal for the above reason and also because all such methods assume exact and rational application of water, something unlikely under the best of circumstances.)

Given such a value for a specific location, it has been assumed that 2.5 times that amount of water per unit area must be removed from river or reservoir in order to meet all other losses and demands and to satisfy the plants. Dunne and Leopold (1978, p. 162), suggest doubling the amount of water needed as a rule of thumb compensating "nonproductive" uses.[2] This value, however, is thought to be too low given the conditions anticipated throughout the GAP region.[3] The inexperience of the users, the extreme length and complexity of the canal delivery system, and perhaps even the inequalities in land ownership found within the area dictate less efficiency of water use.

Given the above, the following relationships are made:

- Total water removed from reservoir or river = 2.5 (evapotranspiration less natural water supply during growing season)
- Return Flow = 0.35 (total water removed)
- Water loss (nonproductive use) = Total water removed less return flow, less water needed to supply irrigation deficit[4]

2. A complete discussion of this question and related matters is found in U.S. Department of Agriculture, Irrigation water requirements (1970, 88).

3. This would be consistent with observations made in southwestern Turkey by this author (Kolars et al. 1983).

4. The definition and use in this study of both "water loss" and "return flow" requires further discussion. A school of thought exists that essentially all the water removed from a river or reservoir which does not directly satisfy actual

- System depletion (total unreturned for all purposes) = water loss plus water needed for irrigation

The Use of the Euphrates in Turkey

Map 10.1 and Map 10.2 show the Euphrates River system in Turkey in its entirety. Table 10.3 traces developments along the river from its headwaters to its debouchement into Syria, as well as related irrigation projects on streams which flow first into Syria before joining the Euphrates. (In the case of the Balik River west of Karkamis, this stream, while not part of the Euphrates drainage system, is shown because of its involvement in the supply of water to the Aleppo area downstream.) The numerous minor projects detailed in Table 10.3 are summarized in Table 10.5. Irrigation areas are shown on Map 10.3 and reservoirs on Map 10.4.

The Keban Dam and reservoir and the smaller projects upstream from that site were among the first developments to be completed on the Turkish Euphrates. Irrigated fields, while developed at an early stage in this area, are of relatively little importance compared to the hydroelectric power plants (HEPP) found here. At this writing, approximately 35,000 ha are under irrigation with

---

evapotranspiration (AE) needs of the crops in question will find its way back into the river system. By this reasoning, such water does not diminish the total remaining, however polluted it may be. Thomas Stouffer (personal communication, Oct 1990) cites conditions in the Nile valley where this is apparently the case. The strict linear character of that system with the narrow confines of the entrenched valley argue for such an interpretation. In the case of the Euphrates and GAP, with its widespread irrigated lands (existing and planned), the size and complexity of the delivery systems (cited elsewhere in this study), and the use of such waters during months of greatest PE, it is assumed that numerous opportunities will exist for evaporaton and transpiration which will neither sustain crops nor allow *all* water unused for crop AE to return to the main stream. Thus the proportions selected for this analysis (row 1, Table 10.4) suggest 25 percent of the water removed from reservoirs constitutes such absolute water loss, 35 percent return flow, and 40 percent actual evapotranspiration. Obviously, any overestimation of water loss would add to the return flow, thus somewhat alleviating the depletion of downstream flow. Only time and the actual implementaton of GAP can answer the questions which this raises. We have chosen a less optimistic view of the situation preferring that surprises, if they come, be positive ones.

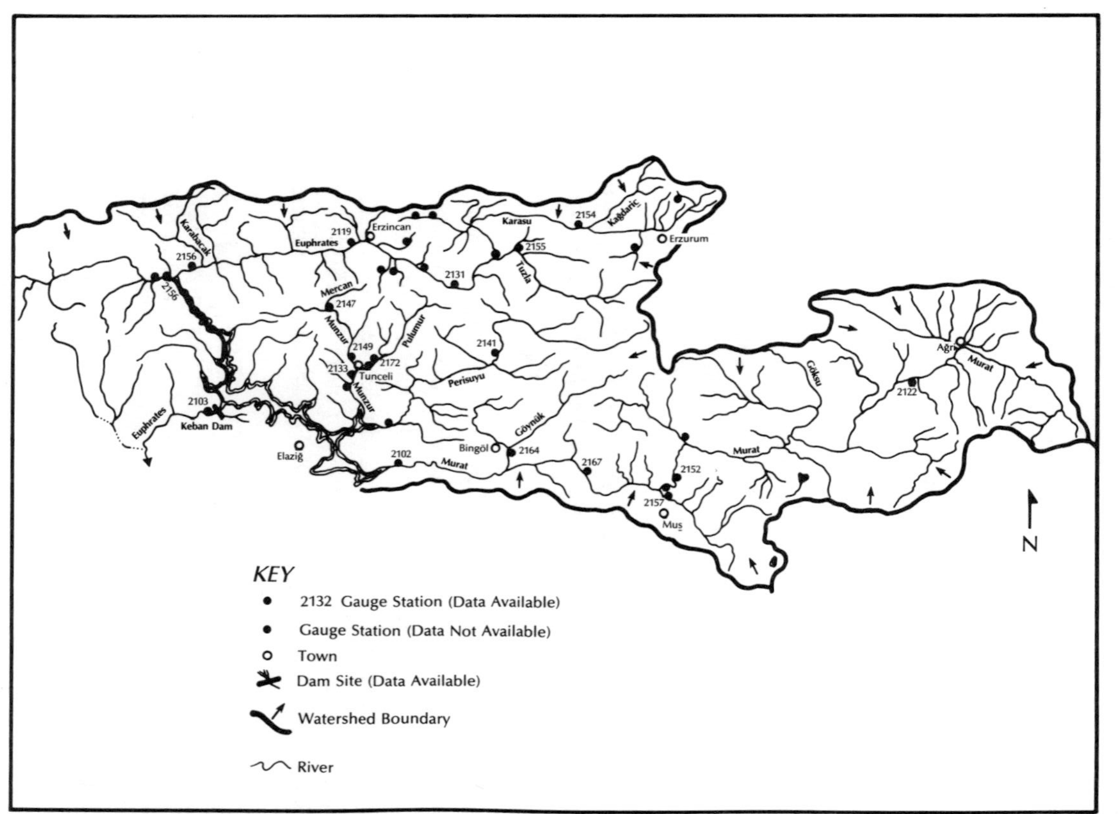

Map 10.1. Headwaters of the Euphrates River.

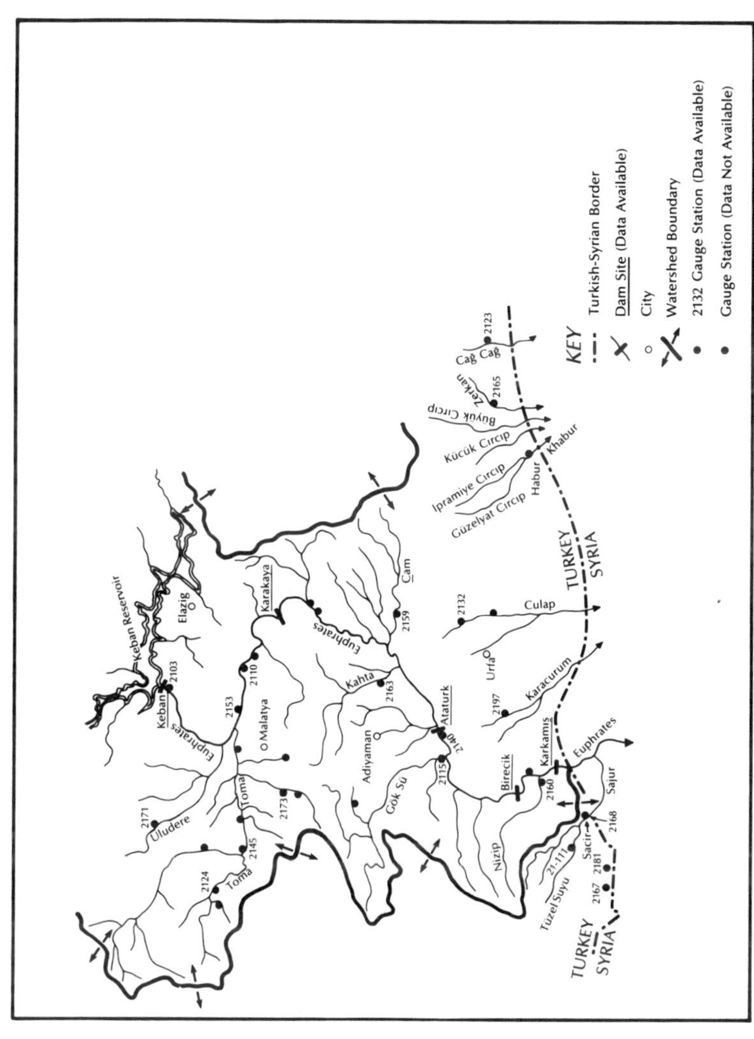

Map 10.2. Hydrography of the Euphrates (Firat) below Keban (with gauging stations). Source: *1982 Water Year Discharges*, Ankara: Elektrik Isl. Etud Id (1958); and GAP (1980).

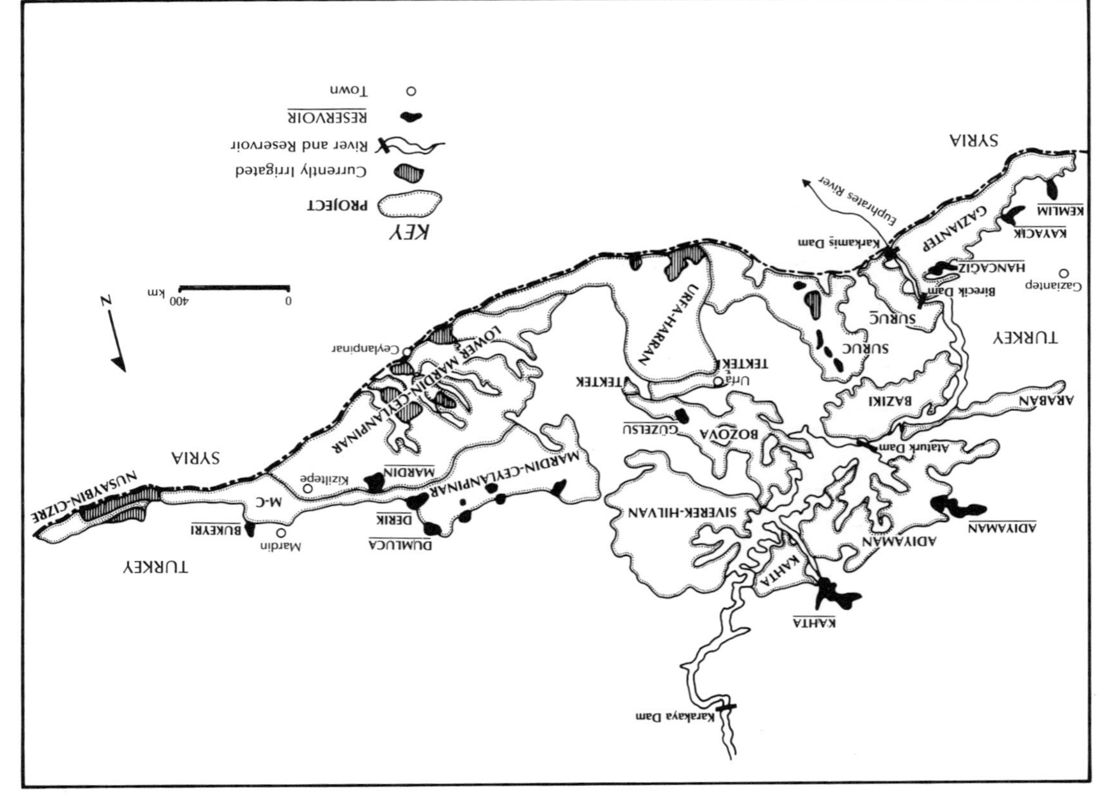

Map 10.3. GAP project areas of the Lower Euphrates. Source: GAP (1980).

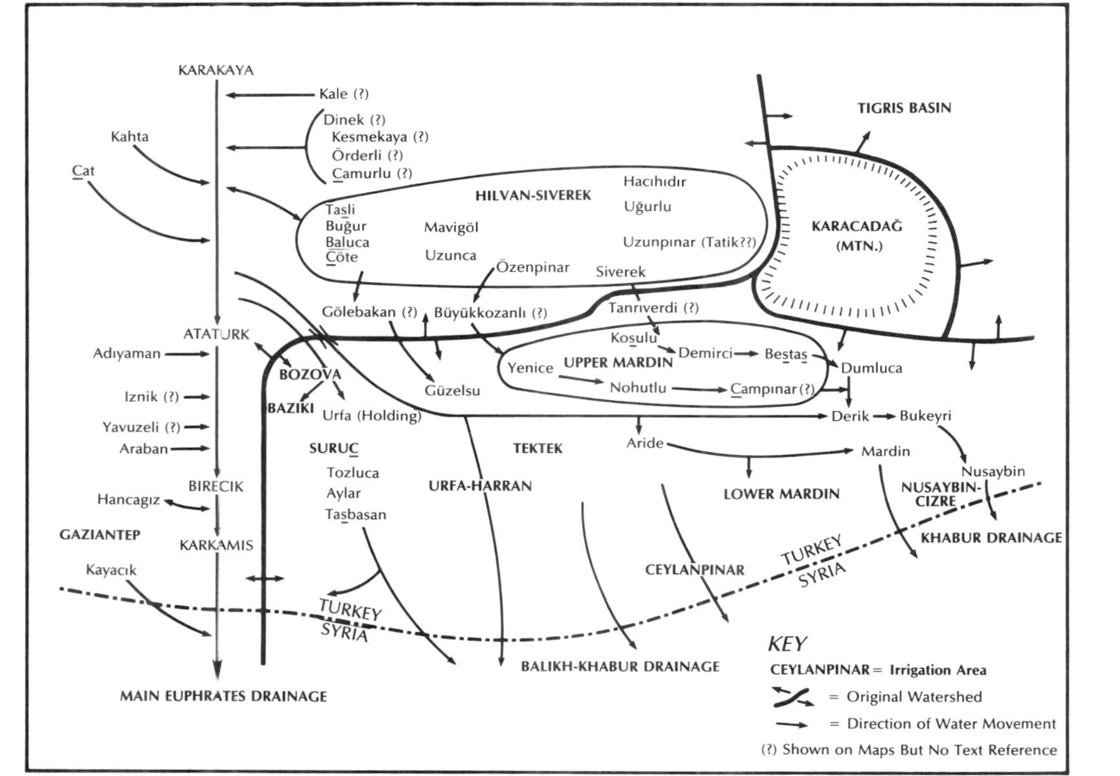

Map 10.4. Named reservoirs of the Southeast Anatolia Project: Karakaya and downstream (approximate locations). Source: GAP (1980) and DSI (1984a).

perhaps 58,231 ha scheduled for about the year 2000. At that time, depletion of river flow, after return flow has taken place, will be 1,431 Mm^3.

Downstream from the Keban as far as the Karakaya Dam is a second section of the developments scheduled by Turkey. At the present time, there is apparently no irrigated farmland, but by the year 2000 about 42,000 ha are scheduled. As noted earlier, the Karakaya reservoir began filling in June 1986. When full, that reservoir may lose as much as 435 Mm^3 of water from evaporation annually. By the year 2000, total depletion of river water for this section should be about 782 Mm^3.

The area between the Karakaya and the Ataturk Dams is by far the most complex and ambitious part of GAP. When fully completed after the year 2000—and if stated GAP goals are met—370, 911 ha will drain into the Euphrates above the Ataturk Dam. Of this amount, 220,511 ha will enter from projects on the right bank (Cat, Adiyaman/Kahta) and the remainder (150,4000 ha) from the Siverek-Hilvan area on the left bank. (It should be noted that this area, though upstream from the Ataturk Dam, is considered as part of the Lower Euphrates project described below.)

The Lower Euphrates Project, which is the core of the GAP, is based upon the Ataturk Dam and its vast reservoir. Eight different irrigation projects totalling 1,148,511 ha are projected for completion sometime after the year 2000. A tentative schedule of when these are expected to come on line is found at the bottom of Table 10.3 and a more detailed account in Table 11.3 and Figure 11.1.

In addition to the 370,911 ha in the above paragraph, 777,600 ha will be irrigated on the southern slopes of the Anti-Taurus Mountains and the plains stretching to the Syrian border. Of this large area, runoff from 378,800 ha will reach the Culap/Balikh system and that from 398,000 ha will flow into the Khabur by way of its many northern tributaries (Table 10.5). At its fullest, the Ataturk Reservoir may lose as much as 1,470 Mm^3 annually to evaporation, and, sometime after the year 2000, depletion of the river from evaporation, water loss, and evapotranspiration might reach the astonishing amount of 13,437 Mm^3 along this section of the stream.

(At this point it seems necessary to pause and emphasize the care with which these figures have been estimated and also to point out that these values represent a complete realization of the project's many features, an eventuality that seems less likely to happen as the magnitude of the venture becomes apparent.)

This depletion will be paralleled by a return flow of 6,461 Mm^3, of which roughly one-third will return to the reservoir and the remaining 4,673 Mm^3 will flow into the Balikh and Khabur systems in Syria.

Downstream from the Ataturk Dam is found the Euphrates Border Project. This includes the Birecik and Karkamis Dams, both of which are intended to generate large amounts of electricity. In addition to hydropower, 101,573 ha are scheduled for irrigation largely from Lake Birecik and the Araban, Hancagiz, and Kayacik reservoirs. Return flow in this case will be about 583 Mm^3 and total depletion for this section, by the year 2000, about 541 Mm^3 rising to 1,257 Mm^3 sometime after that.

In sum, these ambitious plans foresee a region which sometime after the year 2000 will have 1,350,243 ha of irrigated land. Return flow from that land will total 7,408 Mm^3. In the near future—within the next four years—non-recoverable water loss (including some evapotranspiration) will reach 1,976 Mm^3/yr. If all goes according to schedule, this figure should jump to 9,966 Mm^3 by 1995. By the year 2000 it may reach 12,482 Mm^3/yr and sometime after that date might even soar to 16,908 Mm^3. The issue is whether or not this is possible, either technologically, ecologically, or politically.

The Use of the Euphrates in Syria

The first withdrawal of water in Syria downstream from the Turkish border will be on the Syrian portion of the Sajur which enters from the right bank. Little is known about this project, and in any event its magnitude cannot be great.

Next will be the proposed Tishreen Dam, which will create a lake with a volume of 1,300 Mm^3 (*MEED* 1986a) and an area estimated to be about 70 sq km with evaporation loss of 157.5 Mm^3/yr. Immediately downstream, Lake Assad and the Aleppo diversion will remove another 1,570 Mm^3 and 80.2 Mm^3 annually. Lake Assad will also serve five of the six originally proposed irrigation districts. (Rasafah, Area 4, already has been abandoned.) Depletions from these various projects are shown on Table 10.3, Part II. Another dam, the 64 MW Baath, 25 km downstream from Tabqa, was completed in 1986 (Table 8.4). Because of the importance of the Khabur and its development projects it will be treated next as a separate element of this study.

Before that, note should be taken of developments in both Syria and Turkey on the Balikh/Culap and its tributaries. Table 10.5 lists the Turkish projects which will be found in the upper basin of the Balikh/Culap). While irrigation of such magnitude (378,800 ha) would totally dry up any local sources many times over, the major problem facing the lower Balikh in the years ahead would appear to be the problem of managing he return flow which might reach 2,125 Mm^3/yr. Reference is again made to the difficulty in making such estimates and to the variation in quantities depending upon the values chosen (as demonstrated in Table 10.3). Nevertheless, this becomes a major factor in the rational planning of future river use.

Anticipating what will be discussed regarding the Khabur, a maximum estimate shows that Syrian activities will reduce Euphrates flow by 2,100 Mm^3 by 1990 and by perhaps 3,500 Mm^3 in an additional five years. By the year 2000, Syria may be in a position to either take (or lose through evaporation from reservoir surfaces) a total of 12,100 Mm^3 annually. As in the Turkish case, reality must rest in a lesser figure—of which more is said in Chapter 11.

A Critical Pressure Point: The Ceylanpinar/Ras al-Ayn Area

The sources of the Khabur River are shown in Map 9.1. The major perennial source of this stream is a giant spring, the Ras al-Ayn, at the town of the same name immediately across the border from Ceylanpinar, Turkey. This perennial spring, which in reality consists of a number of outlets (Table 9.2), is one of the largest in

Table 10.5. Distribution of Irrigated Areas of Lower Euphrates Project by River into Which Return Flow Drains

Region	Area in ha		Withdrawal in Mm^3	Reservoir Evaporation	Return Flow
To mainstream and Lake Ataturk					
Cat	22,091		304.3	—	106.5
Adiyaman/Kahta	160,000		2,204.0	—	771.4
Cankara	38,420		529.1	—	185.2
Hacihidir	3,400		46.9	7.0	16.4
Siverek-Hilvan	147,000		2,024.9	61.2	708.7
	370,911	(32.3%)			1,788.2
To Balikh in Syria					
And thence to Euphrates					
Urfa-Harran	157,000		2,597.7	—	909.2
Tektek	20,000		537.7	—	188.2
Bozova	55,300		889.4	—	320.2
Baziki (Yaylak)	44,900		742.9	—	260.0
Suruc	101,600		1,680.9	—	588.3
	378,800	(33.0%)			2,265.9
To Khabur (via Khabur, Jagh Jagh,					
and other Euphrates Tributaries					
Lower Mardin/					
Ceylanpinar	140,000		2,474.3	—	866.0
Derik-Mardin	192,100		3,318.3	—	1,161.4
Nusaybin-Cizre	47,000		812.0	—	284.4
Dumluca	2,400		33.1	3.6	11.6
Siverek-Hilvan	17,300		238.3	6.8	83.4
	398,800	(34.7%)			2,406.6
To Syria (Balikh and Khabur combined)					
	777,600	(67.7%)			
Total (Turkey and Syria)					
	1,148,511	(100%)			

Source: DSI (1980); computations by Kolars.

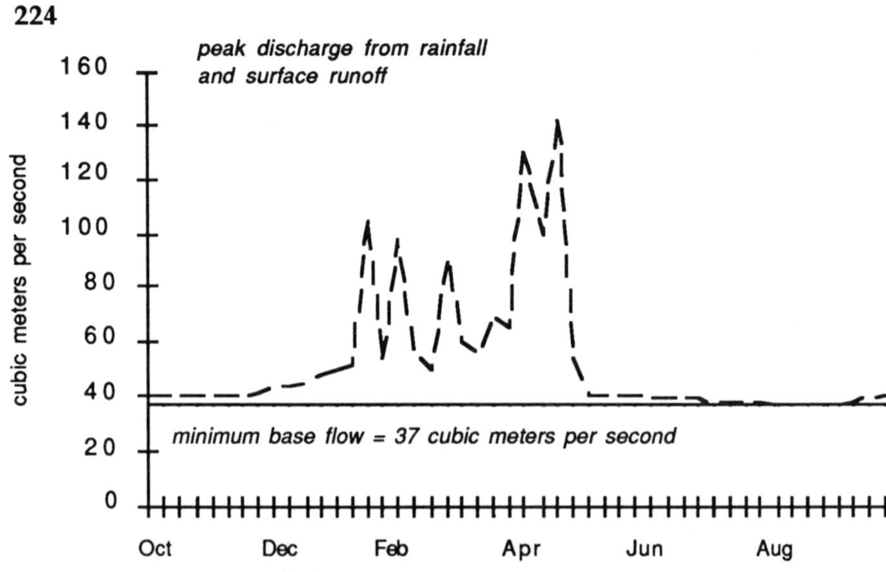

Figure 10.2. Stream flow of the Khabur River at Suwar, Syria, 1932–1933. Source: Dubertret and Weulerse (1940, 62, Figure 57).

the world. Additional water is added to the river by seasonal surface flows from Turkey in the late winter and early spring (Figure 10.2). Other smaller streams also contribute lesser amounts of water to the Khabur. These come from a combination of smaller springs and seasonal runoff. To the east the Jagh Jagh flows from Turkey into Syria as a perennial stream. Farther east and somewhat south is a large marsh, the Radd, which impounds significant quantities of water, much of which is lost through evapotranspiration. The other streams are seasonal in character.

The perennial flow of these streams, with few exceptions, stops just south of the Turkish border. This is the result of a diplomatic and technological coincidence. When the extension of the so-called Berlin-to-Baghdad Railroad was constructed across this territory, the tracks were located far enough up each stream to avoid the expensive bridging of year-round stream flow. Subsequently, when the Turkish-Syrian border was drawn following World War I, the railroad was included in Turkish territory, but so close does the border come to the tracks that in many places one actually steps out of the south side of the train onto Syrian soil. An unforeseen result of all this was that, while the perennial streams and springs feeding the Khabur are in Syrian territory, a large portion of the catchments and aquifers for such springs and streams are located under Turkish administration.

The Ras al-Ayn spring flows at a nearly unvarying rate of 35 m^3 to 40 m^3/s. (It should be noted that the figure "40" in this case represents a real estimated value and not the Middle Eastern "forty.") Figure 10.3 shows this base flow for the Khabur downstream near Suwar and is plotted as a more conservative 37 m^3/s. Winter and spring rains create surface runoff, which begins in January and peaks sometime in April. Spring floods would thus provide an important part of the reservoir storage planned for Syria on the Khabur. At the same time, base flow represents a significant part of the system. The karstic waters of the Ras al-Ayn derive from the aquifer which is located largely across the Turkish border to the north. One account of this recharge area describes it as "7,500 sq km" (UNDTCT 1982), although estimates made for the present study (Table 9.1) are somewhat larger—10,025 sq km. Water-bearing strata dip southward from Turkey into Syria, reaching the surface at Ras Al-Ayn and producing enough head for natural or artesian flow of the waters. Turkish surveys list two areas of underground water availability in the Mardin-Ceylanpinar district: that surrounding Ceylanpinar and another near Mardin-Kiziltepe. The latter is relatively insignificant, having an estimated 13 Mm^3/yr of water recharge, but the former is said to contain a rechargeable supply of 852 Mm^3/yr available for pumping (GAP 1980, III–20). Figure 10.3 indicates that, if all recharge of the Ras al-Ayn spring were to cease, the spring would exhaust its stored supply of water in approximately four years (graph line q), although the unvarying rate of spring flow suggests a much larger fund of stored water.

Two main sources of water ultimately provide for the Mardin-Ceylanpinar/Ras al-Ayn-Jezireh combined region. These are precipitation over the watershed which occurs in the winter and early spring and which declines from 1,306 mm/yr at Lice in the north to 333 mm/yr at Ceylanpinar and to less than 200 mm/yr at Deir

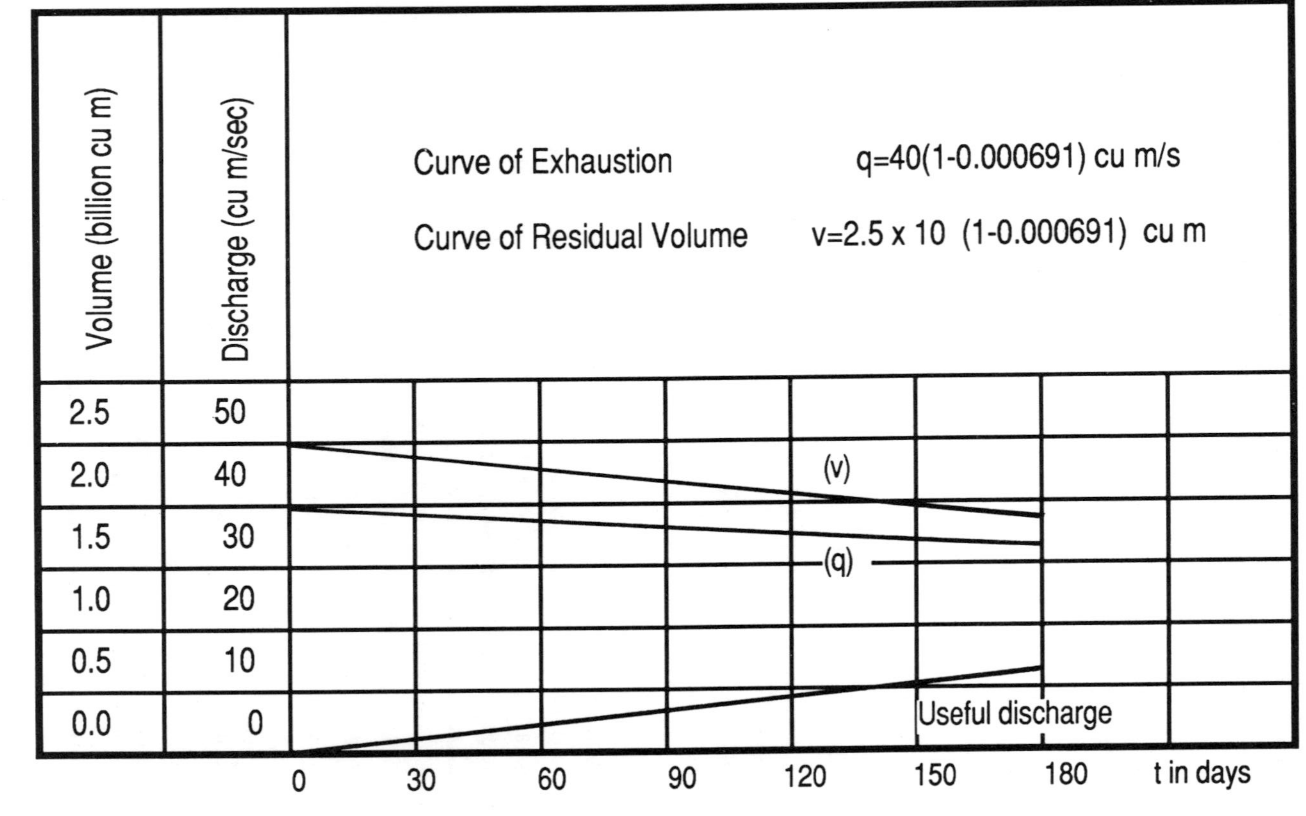

Figure 10.3. Ras al-Ayn exhaustion time. Source: Abd-El-Al (ca 1965, 73).

ez-Zor (Map 9.1). This both provides surface runoff and recharges the underlying aquifers. A second source of water will be that brought into the region from the Ataturk Reservoir. While this water's ultimate source is precipitation farther up the Euphrates River, it is assumed here that such supplies can and will be provided as needed and will be independent of local variation in precipitation at Ceylanpinar.

Seasonal runoff will be partially stored in local reservoirs such as those at Mardin, Aride, and Derik. Another part will flow downstream into Syria, shown in Figure 10.2 as the peak spring flow. Evaporation from these reservoirs will represent a net loss from the system; seepage from them into the aquifer will help to recharge losses from planned pumping. Locally stored waters, as well as water from the main canals leading from the Ataturk Reservoir, will irrigate fields. Additional fields will be served by water pumped from the local aquifer. Evapotranspiration from fields will represent a net loss to the system. Infiltration will partially recharge the aquifer and, in addition, considerable quantities of runoff will move downstream into Syria. In the latter country, plans are underway to irrigate as much as 137,900 ha of land in the upper Khabur basin. The complexity of this situation is such that reference is made at this point to Figures 10.4 and 10.5, which diagram those parts of the overall Khabur/Habur system which are quantifiable. (The following numbered statements refer to corresponding numbers on Figure 10.4.)

1. Precipitation is estimated as the average for the Ceylanpinar-Mardin region times the area of the catchment.
2. The yearly fund of water from the Ceylanpinar aquifer is found in GAP (1980) II-6.
3. The Mardin-Kiziltepe fund (GAP 1980, ii-6).
4. Water use from pumpage is based on GAP plans to irrigate 60,000 ha in the immediate vicinity of the State Farm at Ceylanpinar. This amount is computed as 60,000 ha times 11,489 m^3 water need per hectare (as shown in Table 7.4), which equals 689.3 Mm^3/yr.
5. Remaining flow towards Syria does not take infiltration from fields into account and represents the Ceylanpinar fund less the amount pumped.

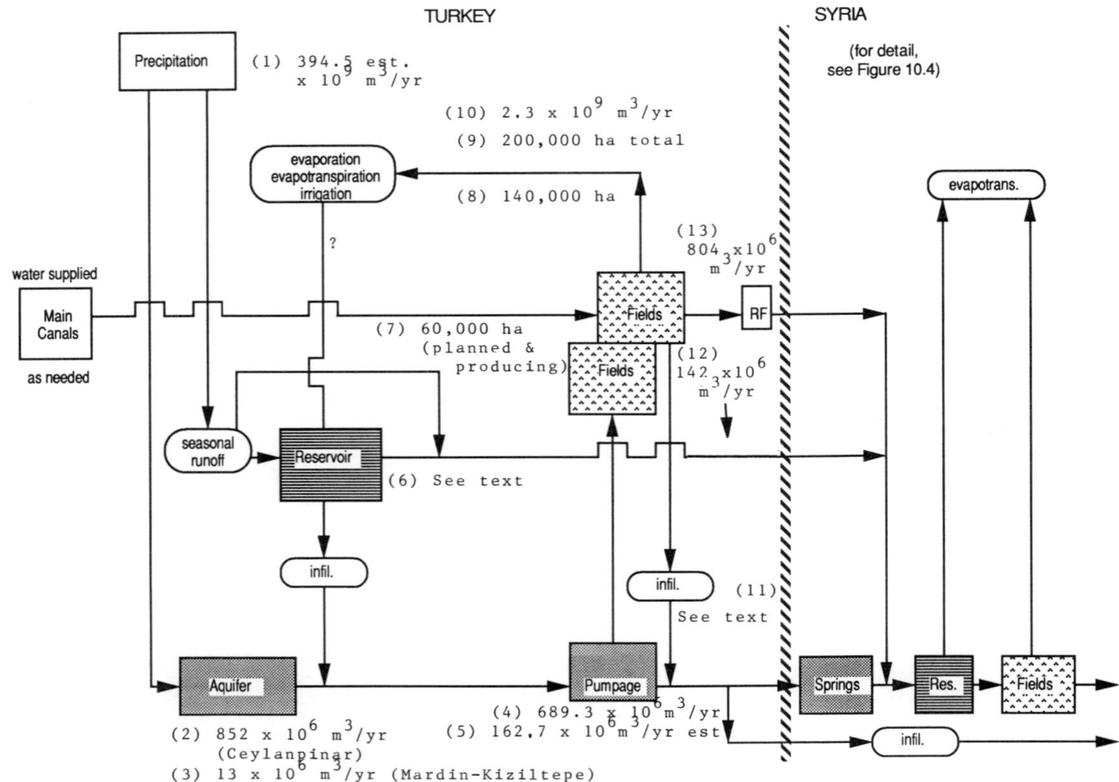

Figure 10.4. Schematic representation of proposed hydrologic relationships in the Ceylanpinar/Ras al-Ayn region (estimated and announced values added where possible).

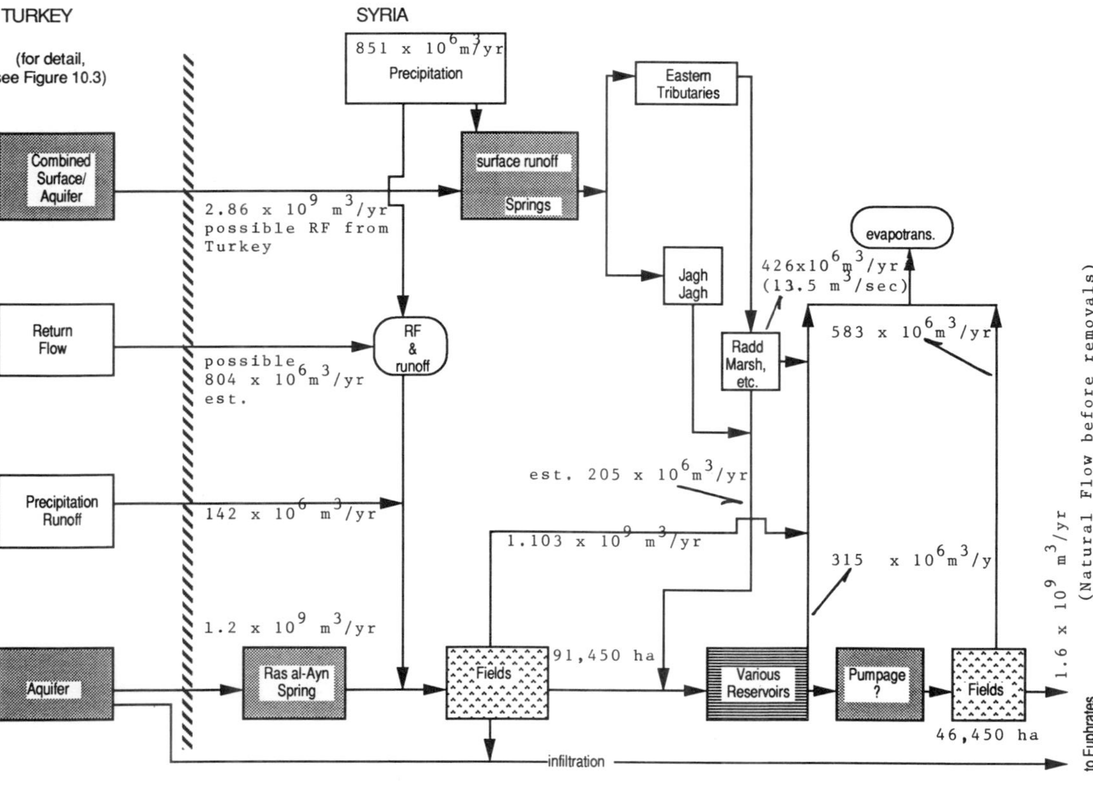

Figure 10.5. Schematic representation of proposed hydrologic relationships in the Ras al-Ayn/Jezirah region.

6. Evaporation from reservoirs will be an overall withdrawal from the Euphrates system but may be replaced locally from the Ataturk Reservoir according to need.
7. Areas of fields receiving pumped water (see 4).
8. Area of fields receiving water from the Urfa Canal—140,000 ha (Table 10.5).
9. Area total: note 7 plus note 8.
10. Water need is that portion of evapotranspiration met by irrigation water supplied by canals or pumping. Natural evapotranspiration and evaporation also represent withdrawals from original precipitation. A value of 11,489 m^3/ha (Table 7.4) may be considered as a conservative value.
11. No infiltration value has been calculated for this diagram.
12. Based on a downstream flow in Syria from this area exclusive of spring flow) of 4.5 m^3/s (Table 9.2).
13. Return flow based on 35 percent of water withdrawn for irrigation times 200,000 ha (200,000 times 11,489 = 2297.8 Mm^3 times 0.35 = 804.2 Mm^3 RF.)

In sum, the overall picture given by this diagram is that pumping will reduce the aquifer fund by about four-fifths of its annual recharge increment—or about 57 percent of the annual flow of the Ras al-Ayn. Return flow will provide 804 Mm^3/yr plus streamflow or an overall flow of 951 Mm^3 on the western (Khabur) portion of the Khabur system. While this may seem a positive factor in the picture presented, water quality of the return flow is also important because the open channels of the karstic aquifer, as well as surface streams, will serve as poor filters compared to sandy strata. It must also be remembered that, while the Khabur may actually gain water, all of the depletion occurring in this sub-region of the Euphrates basin ultimately reduces downstream flow into Iraq.

Surface and spring water for the Jagh Jagh/Radd tributaries to the east are shown in Figure 9.3 along with downstream uses of the Ras al-Ayn/Khabur western tributaries. Precipitation is estimated to add 851 Mm^3 to the eastern area of the Jezirah in Syria.

Of all the eastern tributaries only the Jagh Jagh has perennial flow, and some of the water of this stream has been used in Turkey above Nusaybin for at least a quarter of a century. Downstream in Syria there is an average flow of 205 Mm^3/yr, a relatively small amount compared to the flow of the Khabur to the west. Part of these waters flow directly into the system via the Jagh Jagh; the remainder are filtered through the Radd Marshes where some 425 Mm^3/yr are lost to evapotranspiration (Table 9.2). Total irrigation water needs for the 137,900 ha of field in the entire Khabur system would equal some 1,686 Mm^3/yr (12,226 m^3/ha), almost the amount of the Khabur's average annual flow.

Also at issue at this point is the question of return flow into the eastern tributaries of the Khabur system. Depletion of the existing systems there is less of a question than the one raised concerning the Ras al-Ayn aquifer. On the other hand, as much as 258,800 ha of additional irrigation may be implemented in the eastern portions of the Mardin-Ceylanpinar region. Return flow from these fields would be—at 35 percent of the total water involved—1,538 Mm^3/yr. If this were actually to take place, the entire ecology of the downstream area might be drastically altered. Moreover the question of water quality (addressed below) is again a major issue. Water loss from the reservoirs planned along the Khabur is estimated to approach 333 Mm^3/yr.

Analysis of these data indicates that water loss, as stated, will possibly exceed the annual flow of the Khabur, particularly if evaporation losses from reservoirs take place. Groundwater can supply some of the needed water a long as this source is not seriously depleted in the Turkish catchment area. On the other hand, a significant amount of return flow should find its way downstream from Turkey. If this water is of suitable quality, the immediate crisis of competition for a limited resource may be averted, but only at an ultimate downstream cost through diminution of the total system beginning back at Lake Ataturk. The system closes upon itself at Deir ez-Zor. Downstream returns via the Khabur and Balikh are simply upstream removals less evapotranspiration and evaporation and system inefficiency losses. The overall result will likely be a decrease of flow and increases in impurities.

"Natural Flow" of the Euphrates

Perhaps the most difficult task of an analysis such as this is attempting to learn what the "natural flow" of a river is when so many humans are manipulating it, measuring it, and using its waters. All such activities take place against a constantly changing natural history of climatic variation. In the case of the Euphrates only the broadest estimates can be made regarding what amount of water the river would have in it if people were to leave it alone.

The true natural volume of flow in a river should equal whatever reasonable measured flow can be learned plus some estimate of the upstream uses and/or nonproductive losses.

Table 3.21 list "Irrigated Land Use in GAP Provinces." Of the provinces listed, two will effect the Balikh and Khabur drainage systems through removals for irrigation.

| Mardin | Khabur system | 22,256 ha irrigated |
| Sanliurfa | Balikh system | 33,694 ha irrigated |

Two more may have irrigated lands which either remove water from the mainstream of the Euphrates or deny water to the mainstream by removing quantities from its tributaries.

| Adiyaman | 11,102 ha irrigated |
| Gaziantep | 14,937 ha irrigated |

This might account for a total of 81,989 ha worth of water removed from the system. At an average of 10,000 m^3/ha (1 m^3/sq m), a figure well within the range used in this study, this would deplete the system by 820 Mm^3/yr. [However, the GAP prospectus (GAP 1980) details only 58,309 ha of irrigated land, using 583 Mm^3/yr, in 1980 (Table 10.6). This approximates the amount of irrigated land in Mardin and Sanliurfa provinces.]

Syria has losses of 1,570 Mm^3/yr from Lake Assad, plus those from another estimated 241,000 ha of irrigated land (see Chapter 8, Irrigation in Syria) of which 100,000 ha may have been added in the years since 1973. This would give a possible depletion of 2,100 Mm^3/yr (based on 140,000 ha, a proportion of the depletion shown in Table 8.8 for 241,000 ha.)

Returning to the conclusions of Chapter 5, Average Annual

Table 10.6. Existing Irrigated Land in the GAP Area, ca 1980

Stream/Reservoir Fed

| Location | Stream | Hectares | | Comments |
		Euphrates	Tributary	
Hacikamil (Siverek)	Cam	470	—	enters mainstream
Nusaybin	Cagcag	—	7,820	enters Khabur in Syria
Ceylanpinar	Habur	—	6,700	enters Khabur in Syria
State Production Farm	Small reservoirs	—	2,186	enters Khabur in Syria
STREAM TOTAL		470	16,706	

Pumped from Aquifers

Location	Est. Reserve	Year Begun	Tributary	Comments
Suruc	47 Mm³/yr	1956	6,900	enters Balikh in Syria
Harran	190 Mm³/yr	1974	15,203	enters Balikh in Syria
Akcakale Soil and Agric. Reform Proj.	—		—	
SUBTOTAL BALIKH			22,103	
Ceylanpinar State Production Farm	852 Mm³/yr	1957	8,850	enters Khabur in Syria
	—		—	
Iki Circiparasi	—	1968-80	10,000	enters Khabur in Syria
Mardin-Kiziltepe	13 Mm³/yr	1956	180	
SUBTOTAL KHABUR			19,030	
PUMPED TOTAL			41,133	

Total Estimated Water Depletion (use plus loss)

Basin	Area in ha	Depletion/ ha (see Table 8.4)	Total Depletion in m³	Return Flow in m³
Mainstream	470	8,856	4,162,320	2,214,313
Khabur	35,736	10,754	384,305,000	244,293,000
Balikh	22,103	10,754	237,695,000	151,197,000
Balikh/Khabur	57,839	—	622,000,000	395,390,000
All Euphrates	58,309	—	626,162,000	397,631,000

Source: GAP (1980, II-4, II-7). See this text for source of computations.

Discharge of the Euphrates River, we might take the 28,400 Mm³/yr discharge at Hit, Iraq, and add to it those portions of the above estimates which fall within the calendar range of the observations. That is, evaporation from Lake Assad is not an issue when considering al-Hadithi's (1979) average value, for the lake had not yet been formed at the time his observations were made. Even using the 140,000 ha of irrigated land in the Syrian Euphrates basin is suspect, for there is strong evidence that most of that irrigation began no earlier than the mid-1950s. We therefore must use flow data covering only the time since that date (through 1973) or 29,800 Mm³/yr (Table 5.8). By the same token we must stretch our analysis to add an increment for Turkey of 820 Mm³/yr, although for want of clearer data this will be done.

The end result of all such speculations is:

Hit, Iraq	29,800 Mm³/yr
Syria	2,100 Mm³/yr
Turkey	820 Mm³/yr
	32,720 Mm³/yr

This total is close to the 33,690 Mm³/yr quoted by Clawson et al. (1971) from Hathaway et al. in 1965. Their discussion is given in full in Appendix B. A further check of these figures is provided by data given in the *1983 Statistical Bulletin with Maps* of the DSI (1984b, table 5-1, p. 26).

Careful estimations of the precipitation and annual runoff within the Euphrates drainage basin indicate that 33,480 Mm³/yr of runoff is available from that area. Their map also includes the flows of the headwaters of the Balikh, Khabur, and Jagh Jagh rivers in Syria as well as the mainstream of the Euphrates. It has been argued that the above streams contribute 7 percent of the total flow of the Euphrates entering Iraq as measured at Hit (Chapter 6, Table 6.1). It has further been argued that, of this amount, approximately 80 percent originates in Turkey (Chapter 9, Tables 9.1 and 9.2). In other words, an additional 5.6 percent of the total originates in Turkey or all but 1.4 percent of the water entering Iraq (see Chapter 9). This means that an additional 480 Mm³/yr should be added to the DSI figure to obtain the total runoff reaching Iraq—a total of 33,960 Mm³/yr.

If these three approximations are averaged, we may arrive at a reasonable average figure within a reasonable range of choices for the natural flow of the river:

DSI	33,960 Mm³/yr
Kolars/Mitchell	32,720 Mm³/yr
Hathaway, et al.	33,690 Mm³/yr
100,370/3 =	33,457 Mm³/yr

Sedimentation and Water Quality

There remains the question of the quality of the water which flows both now and in the future along the length of the Euphrates River. Generally speaking, turbidity or suspended solid load increases with volume, while salinity (dissolved load of cations and anions) increases with diminishing volume. Thus, before considering these elements of water quality, it is necessary to discuss the range of volume of water carried by the Euphrates between extreme flood peaks and extreme low water.

Figure 10.6 shows the variation in mean monthly flow of the Euphrates at Hit, Iraq, for the 49-year period 1924–1925/1972–1973, as recorded in Table 10.1. The maximum monthly flow, which occurred in May 1969, was 5,460 m³/s. The extreme monthly low water, which occurred in August 1961, was 94 m³/s. The average annual flow for the 49 year period was 902 m³/s. Momentary peak and minimum flows do not coincide with extreme monthly and yearly averages. Table 10.7 shows the momentary high and low water values for the above period of time. In 1969 an absolute momentary high value of 7,390 m³/s was reached (month unspecified), and in 1973 an extreme momentary low value of 81 m³/s occurred. (Note that, in the latter case, the low monthly average was in 1961, not 1973.) Dunne and Leopold (1978) suggest that momentary discharges be used for computing the

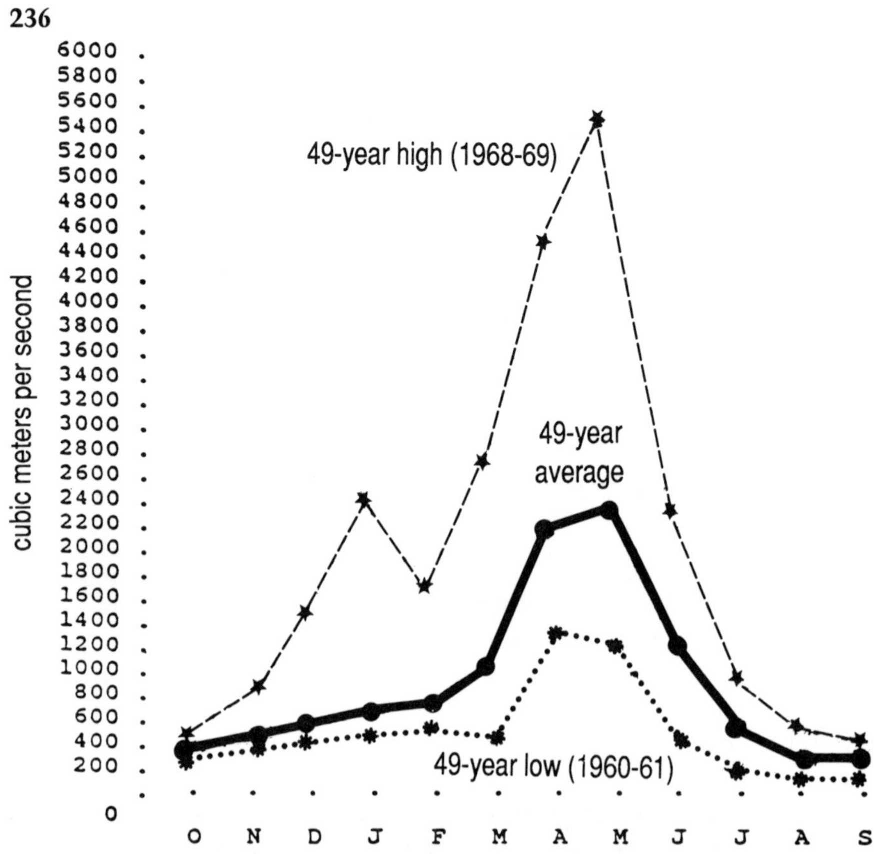

Figure 10.6. Variation in mean monthly flow of the Euphrates at Hit, Iraq (m³/s). Source: al-Hadithi (1978, 225–227, Table 1); computations by Kolars.

frequency of high and low water and the return probabilities of such events.

Figure 10.7 shows the frequency of momentary maximum high water on the Euphrates River. This graph has been prepared using al-Hadithi's data and Gumbel's technique (Dunne and Leopold 1978; al-Hadithi 1979). Extrapolations from this graph should be considered approximate. Nevertheless, by using this, it is possible to obtain an idea of the frequency of flooding on the river assuming that the future record will remain typical of the stream's hydrological past. Thus a flood of the magnitude of that occurring in 1969 will occur, on the average, every 51 years with about 98 percent of all high-water occurrences being less in volume. A maximum flow of 3,525 m^3/s will occur about every two years, while maximum high water of *at least* 4,600 m^3/s will take place every four years and will be exceeded in volume approximately 25 percent of the time.

Figure 10.8 shows the recurrence interval of low-water conditions. Minimum flows of 81 m^3/s or less will occur with a frequency of about 2 percent. On the other hand, low water conditions of 250 m^3/s or less will take place 50 percent of the time, that is, about once every two years. Very few data concerning the suspended load of the Euphrates River are available for this analysis. The river has been described traditionally as extremely turbid during high-water periods. Al-Hadithi (1979) states, "The average sediment load of the Euphrates River at Hit is about 2 kg/m^3," but he does not specify how this figure was reached. He also cites an extreme load of 10 kg/m^3 and a low of from 0.1 to 0.5 kg/m^3. He also indicates that Soviet engineers measured the suspended load "near Deir ez-Zor" sometime prior to 1971 and estimated the annual load as 55 million tonnes per year. Figure 10.9 uses Soviet data given in al-Hadithi for the river at Deir ez-Zor. The direct relationship between the amount transported and the volume of discharge is clearly evident.

It is necessary to point out here that much of this discussion has become moot with the building of the Keban, Karakaya, and Tabqa dams on the river. Each of the reservoirs created by these dams now serves as a settling basin, and, with the addition of still more dams and reservoirs, the river will become less and less turbid. Nor will early estimates of the life of reservoirs remain valid, for the addition of each new settling basin will change and lengthen the life span of those farther downstream.

It should be noted, as shown in Map 10.5, that the river in Syria is incised within the north Syrian upland and has a rather narrow floodplain bordered by bluffs and upland surfaces some 60–80 m higher than the river. The effect there is threefold. First, the easily irrigated land reached by pumping directly from the river is restricted to the floodplain. Second, the water table beneath

Table 10.7. Recorded Flows at Hit, Iraq (in m^3/s)

Part I—Peak (1924–1973)

Year	Peak Flow	m	$T = \dfrac{n+1}{m}$	Year	Peak Flow	m	$T = \dfrac{n+1}{m}$
1969	7390	1	51.00	1964	3548	26	1.96
1968	6654	2	25.50	1936	3450	27	1.89
1967	6072	3	17.00	1965	3422	28	1.82
1929	4980	4	12.75	1926	3320	29	1.76
1963	4816	5	10.20	1937	3320	30	1.70
1972	4810	6	8.50	1928	3240	31	1.65
1954	4730	7	7.29	1935	3200	32	1.59
1948	4670	8	6.38	1949	2950	33	1.55
1940	4660	9	5.67	1947	2900	34	1.50
1952	4610	10	5.10	1959	2770	35	1.46
1953	4540	11	4.64	1955	2600	36	1.42
1944	4530	12	4.25	1970	2550	37	1.38
1938	4500	13	3.92	1945	2510	38	1.34
1966	4484	14	3.64	1958	2480	39	1.31
1971	4435	15	3.40	1951	2470	40	1.28
1956	4430	16	3.19	1962	2224	41	1.24
1957	4420	17	3.50	1933	2170	42	1.21
1941	4220	18	2.83	1924	2120	43	1.19
1960	4080	19	2.68	1973	2055	44	1.16
1942	4040	20	2.55	1927	1850	45	1.13
1943	3900	21	2.43	1925	1750	46	1.11
1939	3850	22	2.32	1961	1732	47	1.09
1946	3750	23	2.22	1934	1730	48	1.06
1950	3690	24	2.13	1932	1630	49	1.04
1931	3630	25	2.04	1930	850	50	1.02

(continued)

Table 10.7. (continued)

Part II—Minimum (1925–1973)

Year	Peak Flow	m	$T = \frac{n+1}{m}$	Year	Peak Flow	m	$T = \frac{n+1}{m}$
1973	81	1	50.00	1960	253	26	1.92
1961	94	2	25.00	1947	261	27	1.85
1970	150	3	16.70	1950	264	28	1.79
1962	153	4	12.50	1956	269	29	1.72
1964	162	5	10.00	1949	273	30	1.67
1925	177	6	8.30	1937	275	31	1.61
1959	194	7	7.10	1952	281	32	1.56
1958	196	8	6.30	1948	281	33	1.52
1927	196	9	5.60	1945	290	34	1.47
1930	201	10	5.00	1938	291	35	1.43
1928	208	11	4.60	1929	298	36	1.39
1934	209	12	4.20	1941	303	37	1.35
1932	213	13	3.90	1946	304	38	1.32
1933	215	14	3.60	1966	304	39	1.28
1965	218	15	3.30	1953	308	40	1.25
1972	224	16	3.10	1943	309	41	1.22
1951	226	17	2.90	1944	330	42	1.19
1926	228	18	2.80	1936	331	43	1.16
1955	228	19	2.60	1954	336	44	1.14
1935	236	20	2.50	1940	343	45	1.11
1942	238	21	2.40	1939	359	46	1.09
1957	238	22	2.30	1969	404	47	1.05
1931	240	23	2.20	1967	408	48	1.04
1963	248	24	2.10	1968	453	49	1.02
1971	251	25	2.00				

Source: al-Hadithi (1978, 228, Table E-2, and 236, Table E-5).

N.B. Peak and minimum momentary flows do not coincide with peak and minimum monthly and/or yearly averages at all times. Nevertheless, Dunne and Leopold (1978) suggest that momentary discharges be used for these computations.

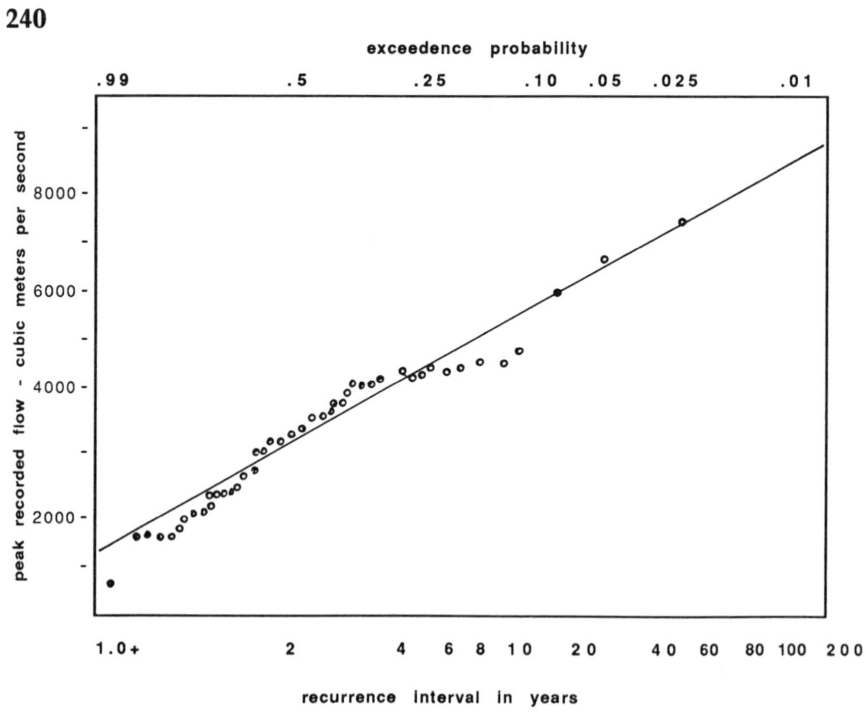

Figure 10.7. Flood frequency of the Euphrates River at Hit, 1925–73. Source: al-Hadithi (1978, 228–231); Dunne and Leopold (1978, 305–309). Note: Some observations not plotted to avoid crowding.

floodplain soil is near the surface with consequent problems of drainage and salination. Third, as reservoirs are put in place along the river in Syria much of the land formerly cultivated by means of small-scale, pumped irrigation will be flooded and new soils at higher elevations have to be utilized.

The outcome of the impounding of the river may be somewhat similar to the problems encountered along the Nile. Less suspended load will increase the velocity of the water in the mainstream with subsequent undercutting of man-made emplacements downstream and the reshaping of the channel in unpredictable ways. (In the absence of up-to-date and/or complete observations, these comments must remain speculative.)

Water quality in terms of the dissolved load is an even more

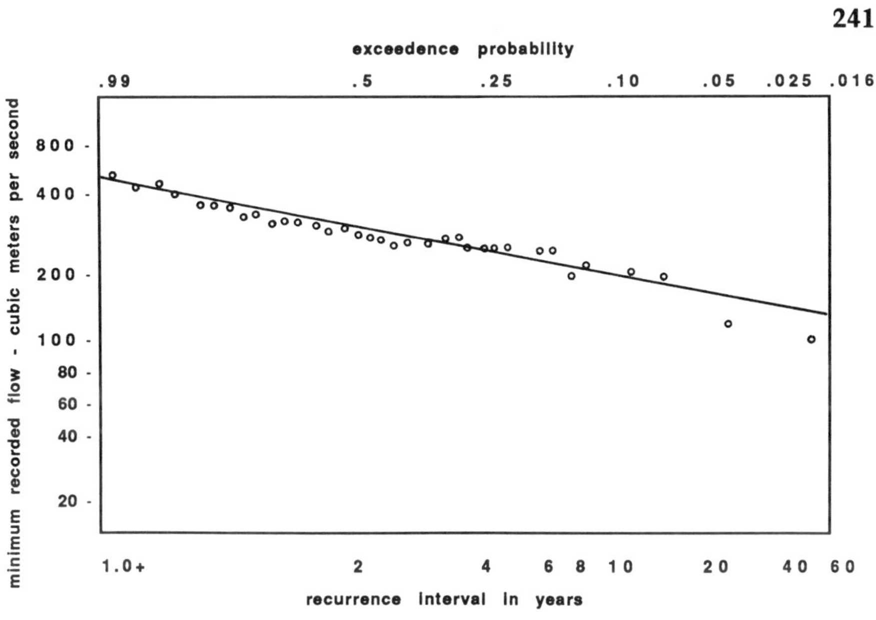

Figure 10.8. Euphrates River at Hit, Iraq, 1925–1973 (low-water frequency curve). Source: al-Hadithi (1978, 236–237). Note: Some observations not plotted to avoid crowding.

important issue than suspended solid load. Figures 10.10 and 10.11 illustrate two characteristics of streams vis-a-vis dissolved load. The first relationship simply stated is: the less water in the stream, the more concentrated will be the dissolved load it carries; the more water in the stream, the more diluted will that load become. That this holds true for the Euphrates is further illustrated by Figure 10.12 and Table 10.8, which plot average monthly water volumes (at Hit, which is assumed to be a surrogate measure of conditions at Deir ez-Zor, where the salinities were measured) against total salinity measured in micromhos/m^3. While the two data sets shown are separated in time and space, it is assumed that the general condition they illustrate will hold true.

Figure 10.11 displays a further relationship common to streams used for multiple irrigation projects. That is, the farther downstream and the more times the water in the stream has been

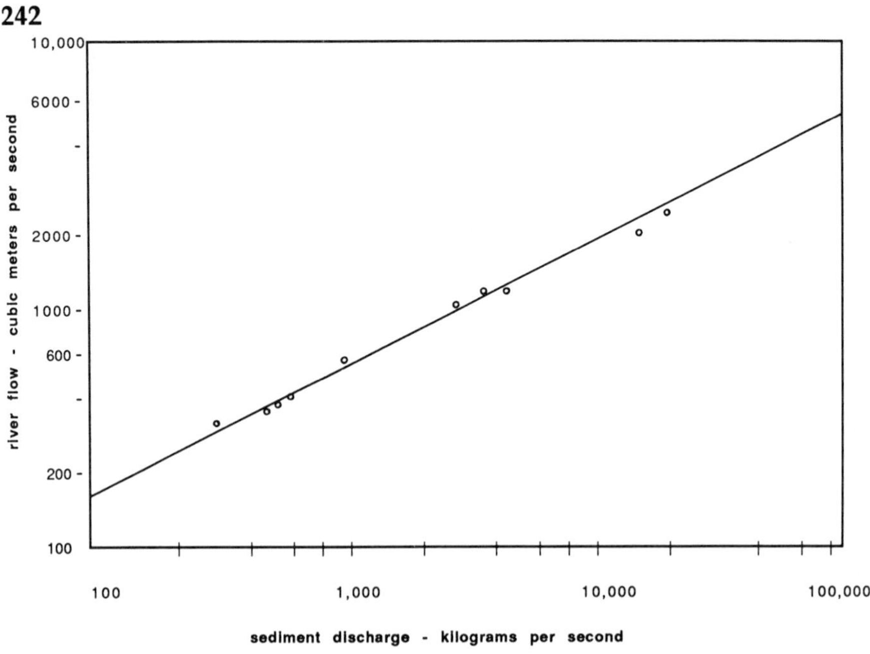

Figure 10.9. Euphrates River sediment discharge at Deir ez-Zor, Syria prior to Keban and Tabqa Dams. Source: After al-Hadithi (1978, 218, Figure D-1, attributed to Hydroproject, *Rawa Hydroelectric Project on the Euphrates River*, Moscow, 1971.

passed through irrigated fields, the more concentrated will its various dissolved salts become. Dunne and Leopold (1978) cite the case of the Colorado River, where the salinity of the river at Lee's Ferry is rising 32.8 mg/l for every 100,000 ha of newly irrigated land. Comparison with the multiple irrigation projects along the Euphrates and its tributaries is obvious. That this holds true is also indicated by the two salinity curves shown on Figure 10.12, where the salinity curve for Tabqa is consistently lower than the curve for Deir ez-Zor farther downstream.

Available values for the salinity of the Euphrates fall between 427 and 760 micromhos/m^3. Figure 10.13 shows the USDA classification of irrigation waters with regard to dissolved salts. Thus the quoted salinities (Tables 10.9 and 10.10) fall within the me-

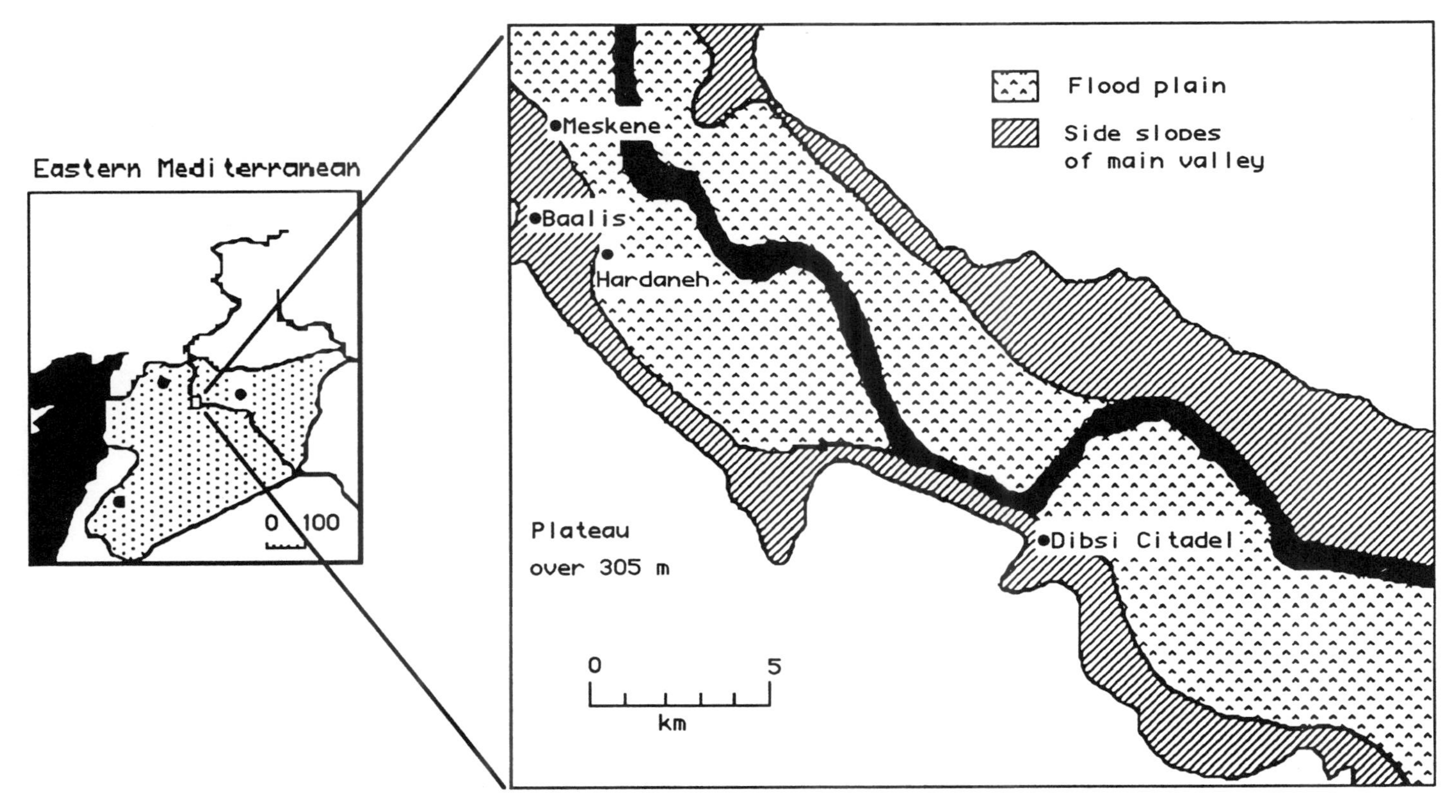

Map 10.5. Valley of the Euphrates River near Meskene, Syria (showing the floodplain and bluffs). Source: Wilkinson (1978). Note: This area now flooded.

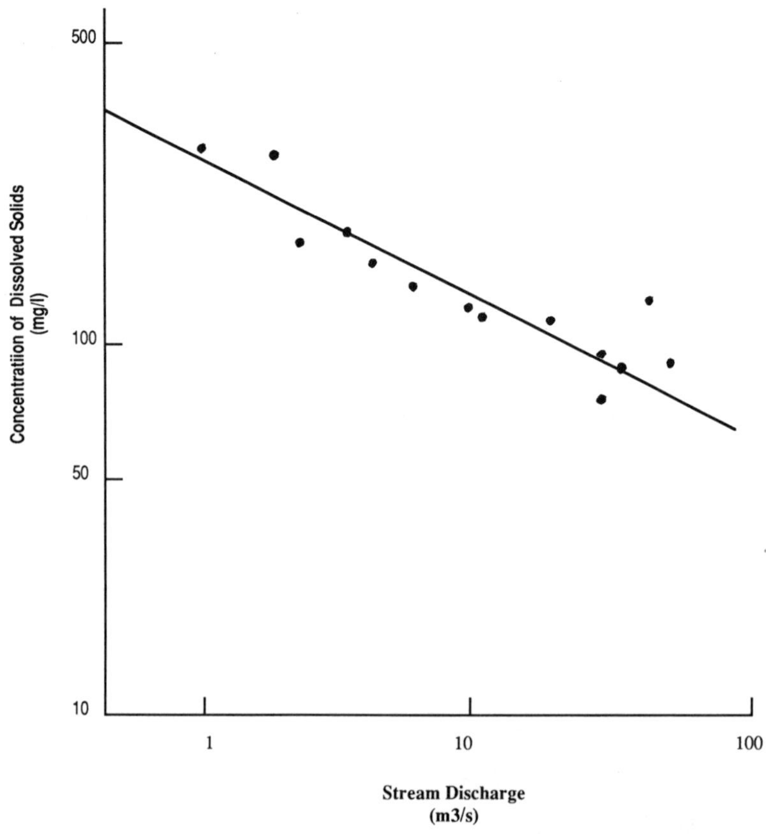

Figure 10.10. Variation of total dissolved solids concentration with stream discharge for the Athi River at Ol Donyo Sabuk, Kenya. Source: Dunne and Leopold (1978). Copyright © 1967 by the American Association for the Advancement of Science. Reprinted by permission of the American Association for the Advancement of Science.

dium hazard range with the exception of October (year unspecified) at Deir ez-Zor. The FAO report for the Jezirah (FAO 1966) states that the rivers of the Jezirah are only slightly salinized (0.27 to 0.72 mg/l) and can be used for irrigation without difficulty. By the same token the two river samples shown in Table 10.10 fall within the USDA classification of C2-S1 of medium salinity and

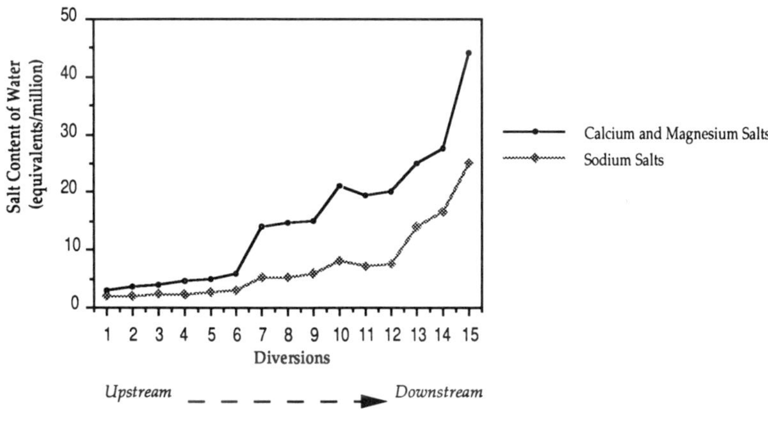

Figure 10.11. Changes in the salt content of the Sevier River, Utah, as a result of repeated diversion for irrigation. Source: Dunne and Leopold (1978, 153) from Thorne and Peterson (1967).

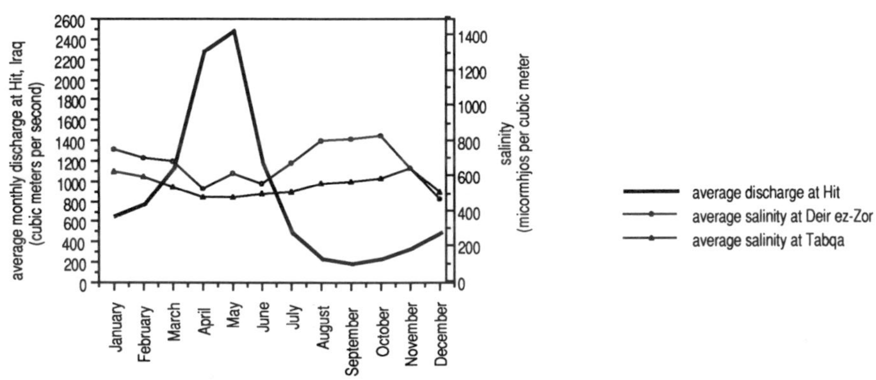

Figure 10.12. Salinity and Average monthly discharge at Hit, Iraq (m^3/s). Source: Raslan and Fardawi (1971, 216); Clawson et al. (1971, Table B-10, 218).

Table 10.8. Salinity at Two Different Locations on the Euphrates River[*]
(micromhos/m^3)

Location	Jan	Feb	Mar	Apr	May	Jun	Jul	Aug	Sep	Oct	Nov	Dec	Average
Tabqa	550	530	475	420	420	430	480	505	525	565	615	450	497
Deir ez-Zor	660	610	600	455	560	480	625	725	735	760	700	480	616

Source: Raslan and Fardawi (1971, 216).

[*]Number of years unspecified; probably a 1-year sample.

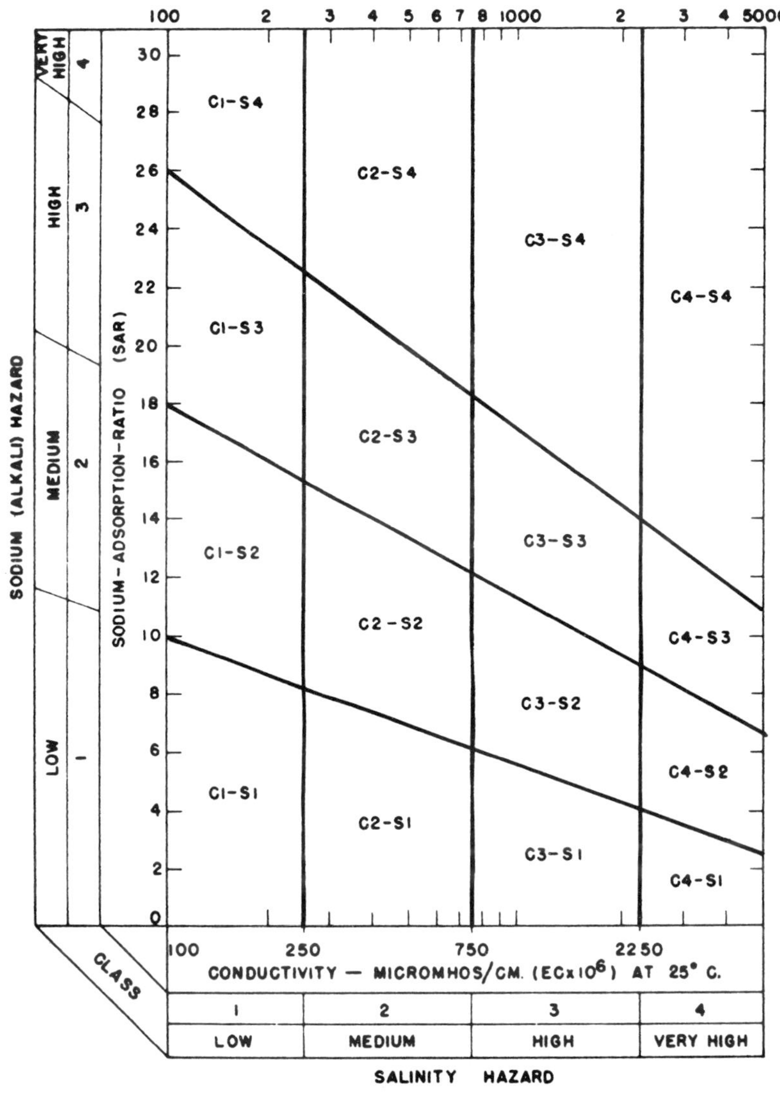

Figure 10.13. USDA classification of irrigation waters. Source: Withers and Vipond (1980, 114) from USDA, "Diagnosis and improvement of saline and alkaline soils," *Agricultural Handbook,* No. 60.

medium sodium hazard. On the other hand, Withers and Vipond (1980, 113) believe that such medium sodium-rich waters should be used only with coarse textured, permeable soils.

There remains the question of how dramatically increased hectarages of irrigated land with subsequent return flow to the tributaries and the mainstream of the Euphrates will effect downstream users. Maps 10.6, 10.7 and 10.8 show the concentration and distribution of the most prevalent dissolved salts in the underground waters of the Jezirah. The general distribution of cations and anions is shown on Map 10.6. Dilute bicarbonates predominate along the Turkish border. This is typical of the good quality of the Jezirah streams at their point of origin in the north. Nevertheless, greater concentrations of bicarbonates occur along the southern border of this zone. Excessive concentrations of chlorides are found in the Radd Marsh (the result of high evapotranspiration) and in the south along the Euphrates River, as well as in the east along the Iraqi border, where temporary seasonal accumulations of water evaporate. Sulfates predominate in areas with lower precipitation and ephemeral streams. While many wells produce water suitable for agriculture, drinking water from these sources is less available.

There is a question relating to the above FAO survey of underground waters of the Jezirah (FAO 1966). The suggestion made in the conclusion of that report is that skillful management of pumped wells would provide the best means of farming in the Syrian Jezirah. However, little subsequent effort seems to have been made to follow that plan, and, instead, the use of surface waters impounded by dams (described elsewhere in this chapter) has predominated. High salinities in a number of wells (see Table 10.9) may account for this change in development priorities, but the question remains unanswered.

An early review of the problem of salinity in the Euphrates Valley of Syria estimates that more than 20,000 ha had already been taken out of production because of high salinity; that, in another 20,000 ha, the yield had been decreased by 50 percent; and, in 60,000 more ha, yield was lowered by 20 percent. This amounted to a total loss of about 70,000 tonnes of cotton per year (Raslan and Fardawi 1971). That this remains a problem is

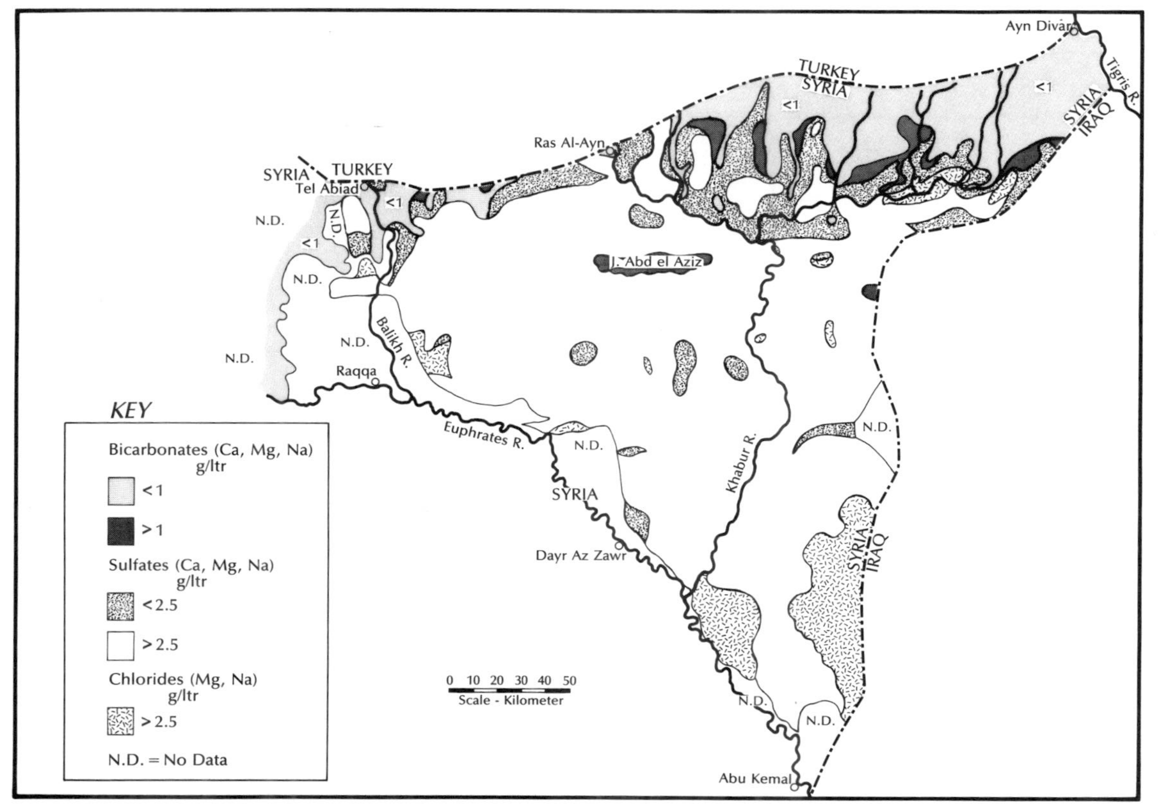

Map 10.6.   Groundwater hydrochemistry of the Jezirah, Syria. Source: FAO (1966, Map 4).

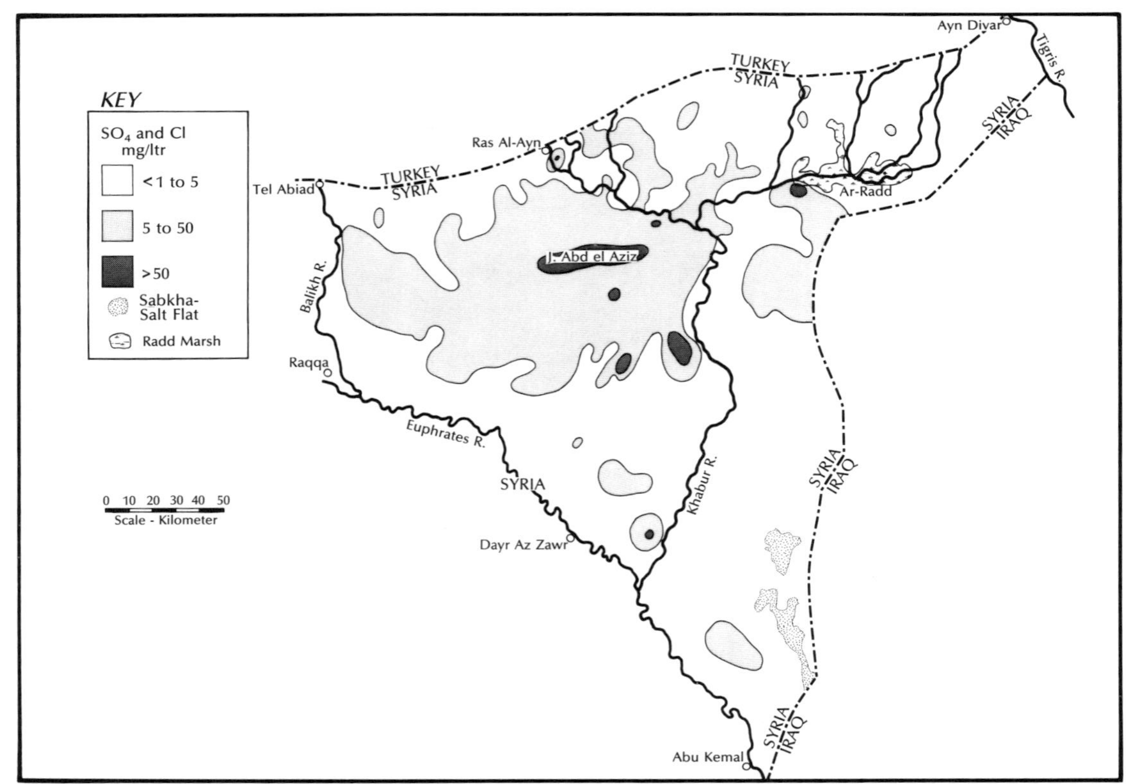

Map 10.7. Concentration of anions in the groundwaters of the Jezirah, Syria. Source: FAO (1966, endpapers, Map 4).

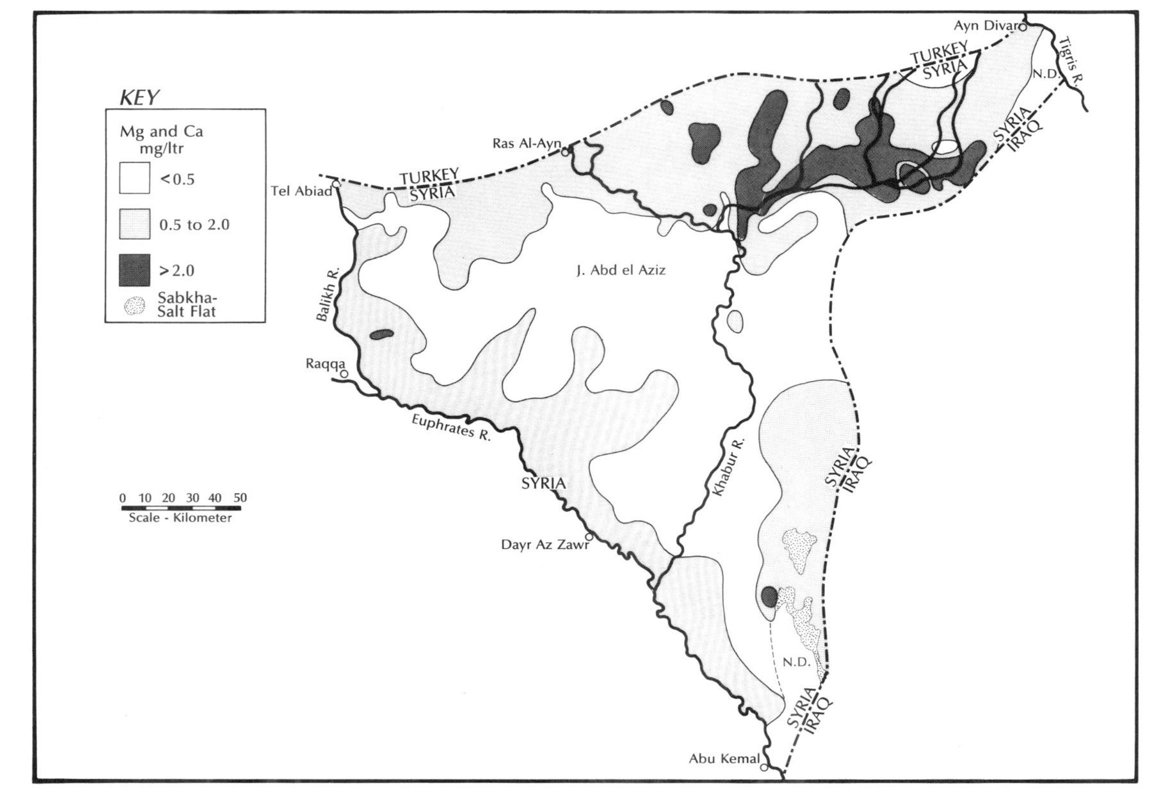

Map 10.8.   Concentration of cations in the groundwaters of the Jezirah, Syria. Source: FAO (1966, endpapers, Map 4).

Table 10.9. Composition and Concentration of Salinity in the Syrian Jezirah

Sample*	EC × 10^3 mmhos/cm	pH	Ca	Mg	K	Na	NH$_4$	CO$_3$	HCO$_3$	Cl$_4$	SO$_4$	NO$_3$	SAR
Euphrates	0.484	7.3	2.90	1.53	0.33	1.87	0.11	—	3.72	0.24	0.63	0.06	1.25
Euphrates	0.427	7.4	0.88	1.72	0.13	2.48	0.11	0.17	2.60	0.92	1.08	0.02	2.07
Well	1.420	7.4	7.80	3.36	0.18	5.65	0.06	0.33	3.64	6.53	3.37	0.02	2.40
Well	12.100	7.1	24.20	58.60	1.03	78.26	0.11	0.25	2.44	62.10	83.20	11.60	12.20
Well	27.923	7.4	18.00	179.20	1.03	230.40	0.33	0.83	8.06	173.80	244.70	0.02	23.20

Source: Raslan and Fardawi (1971, 217).

SAR = Sodium absorption ratio.
*sample locations unspecified except as shown.

indicated by current reports (*MEED* 1986b) that the World Bank is considering loans to Syria to provide for a second stage Lower Syrian Euphrates drainage and irrigation scheme. First stage work, to be completed in mid-1987, has already been financed by the World Bank and entails more than 200,000 ha. Second stage work is intended to reduce salinity and create an effective irrigation network for an additional 120,000 ha of reclaimed land. From these efforts it is clear that excess salinity remains a significant problem in that area.

Turkish data referring to water quality in the GAP area have yet to become available for analysis. The original GAP prospectus (GAP 1980) devotes only a brief, non-detailed commentary to this subject. Ayyildiz et al. (1986) indicate that some drainage problems must be dealt with, particularly in the more southerly portions of the Harran Plain. They state that, given a salinity measure of 400 micromhos/m^3 as an average value for waters in the Urfa-Harran region and an estimated irrigation need of 1,148 mm in order to produce a cotton crop,[5] such waters will deliver approximately three tons of salt per year per hectare (261 mg/l) which must be carefully leached away. This inevitably implies that such materials would be transported farther downstream.

The above brief review of water quality in the Euphrates basin assumes relatively few problems will occur for Turkish use of the water either from sedimentation or salinity. In Syria, however, problems are already occurring along the mainstream. That these will be even more serious in the future becomes evident when a sequential water budget of the combined Turkish and Syrian river system is made. Figure 10.14 and Table 10.11 depict the elements and values in such an accounting. All data are drawn from this report and are summarized in Table 10.3. What this schemata attempts to show is how demands made upon the river's water resources will vary sequentially with withdrawals (w_t, w_s) and return flows (RF_t, RF_s). Evaporation from reservoir surfaces will also take its toll (E_t, E_s). (Other terms used in the diagram are "Aleppo" for the water withdrawn to provide the city of Aleppo,

---

5. This is apparently based on PE values rather than on the water deficit shown by a computed water balance, which would be less. (See Table 8.3)

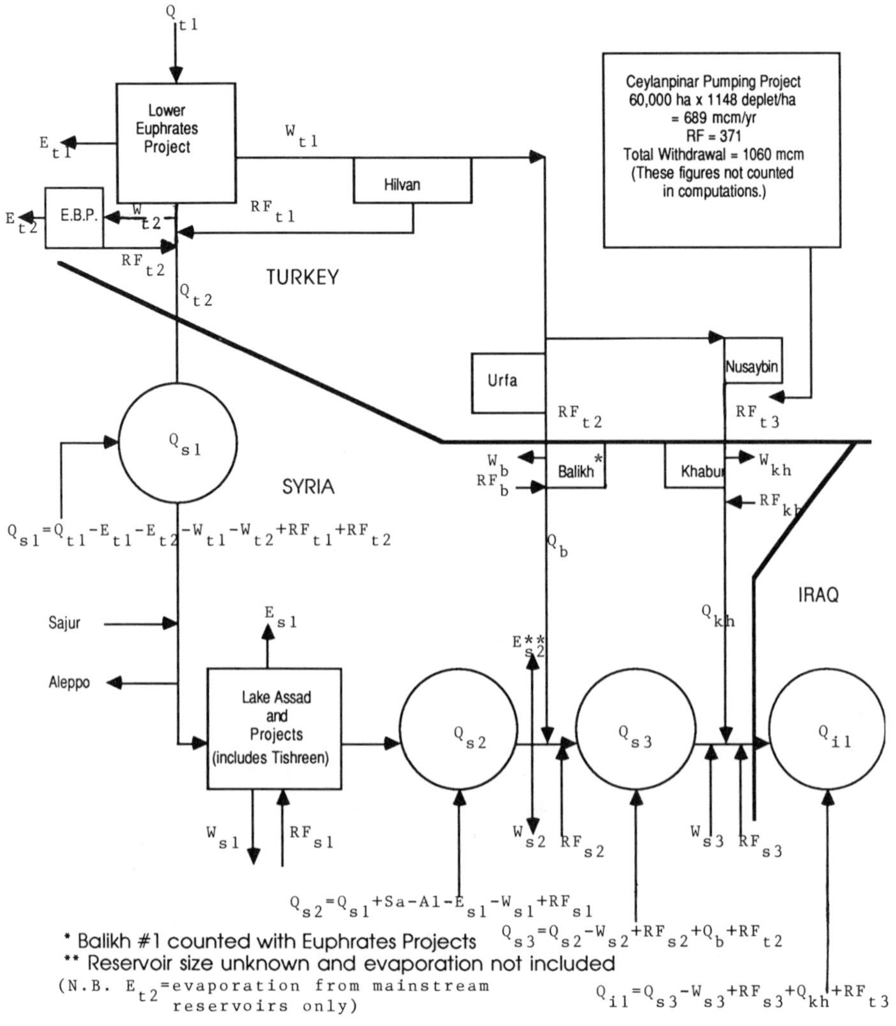

Figure 10.14. Sequential water budget of the Euphrates River, ca 2000+.

Table 10.10. Iraq's Projected Share of Euphrates Water, 1986–2000+
(in Mm³, worst case scenario)

	1986–1990	1990–1995	1995–2000	2000+
Estimated "Natural Flow" entering Iraq	33,460	33,460	33,460	33,460
Combined Turkish and Syrian Use of Water	4,109	13,460	24,562	28,987
Share Remaining for Iraq	29,351	20,000	8,898	4,473

"Sajur" for the input of the Sajur in Syria, and "E.B.P." for Euphrates Border Project.)

The results of this preliminary bookkeeping vis-a-vis the river's waters show that after the year 2000, *if all the proposed projects described in this report were to actually be put in place,* Syria would receive 9,442 Mm³ (299 m³/s annual average) from Turkey at the point where the river crosses the border. Initial withdrawals and return flows in Syria would reduce river volume just below the Tabqa Dam to a mere 1,067 Mm³ (34 m³/s annual average). Additions from the Balikh plus return flows from Turkey (originating from Lake Ataturk, but brought across country to the large southeastern irrigation projects) would increase mainstream flow to 2,850 Mm³ (90 m³/s annual average). Similar inputs from the Khabur farther downstream would mean that Iraq might expect from 4,086 to as little as 3,397 Mm³ (130 m³/s to 108 m³/s) annual average.

It was assumed, in making these computations, that reservoirs in Turkey would reduce or eliminate extreme variation in the flow of the stream between flood peaks and drought deficiencies both on an annual and long-term basis. Nevertheless, severe diminution of flow would result from human activities, and all return flows would be heavily salinized. Thus it can be reasonably predicted that the water entering Iraq under such conditions

Table 10.11. Water Budget: "Planned Scenario" for Euphrates River
From Headwaters to the Iraqi Border, 2000+
(E = evaporation; W = total withdrawals; RF = return flow)

Section of River		Amount	
This beginning amount equals the average "natural flow" at Hit less 7% added in Syria.		31,115	Mm^3/yr.
Karakaya Dam and upstream	E −	1,488	
		29,627	
	W −	1,116	
		28,511	
	RF+	391	
At boundary of Lower Euphrates Project Q_{t1}		28,902	
Ataturk Reservoir	E −	1,471	
		27,431	
Small Reservoirs	E −	77	
		27,354	
To. L.E.P. irrigation	W −	18,434	
		8,920	
From L.E.P. irrigation (Table 10.5)	RF+	1,788	
		10,708	
Euphrates Border Project	E −	184	
		10,524	
	W −	1,665	
		8,859	
	RF+	583	
At Turkish/Syrian border Q_{s1}		9,442	= 299 m^3/yr
To Syria via Urfa Tunnel—See Table 10.5			
Small reservoirs	E −	10	(Omitted below)
To Balikh via Urfa, etc.	W −	6,449	(Entered below)
To Balikh via Urfa, etc.	RF +	2,266	(Entered below)
To Khabur via Habur, etc.	W −	6,876	(Entered below)
To Khabur via Habur, etc.	RF +	2,407	(Entered below)

(continued)

Table 10.11. (continued)

Section of River		Amount	
At Syrian Border Q_{s1} (see above)		9,442	
Tishreen Reservoir and Lake Assad	E −	1,727	
(also: Aleppo −80.2 and Sajur R. +80.8)		7,715	
Maskanah, Balikh, Lower Valley	W −	10,228	
		− 2,513	
	RF +	3,579	
Q_{s2}		+ 1,066	
Mayadin Plain	W −	1,036	
		29	
	RF +	363	
		392	
Balikh (stream flow)	+	190	
Harran/Urfa (Turkish)	RF +	2,266	
Q_{s3}		2,848	
Lower Khabur	W −	1,942	
		906	
	RF +	679	
		1,585	
carried forward		1,585	
Khabur (stream flow after extractions)		94	
		1,679	
Ceylanpinar/Nusaybin (Turkish)	RF +	2,407	
Q_{i1} (N.B. This total differs slightly from that given in Figure 10.14 due to rounding errors.)		4,086	= 130 m^3

Revised Estimate excluding Mayadin Plain and Lower Khabur

Section of River		Amount	
Below Maskanah, Balikh, Lower Valley (see above)		1,066	
Balikh (stream flow)		190	
Harran/Urfa (Turkish)	RF +	2,266	
		3,522	
Khabur (stream flow after extractions)		94	
Ceylanpinar/Nusaybin (Turkish)	RF +	2,407	
Q_{i1}		6,023	= 191 m^3

would be of little or no use save for flushing the main channel of the stream.

Conclusion

"Total Depletions to the Iraqi Border" concludes Table 10.4. Given the caveats expressed throughout this analysis, the picture revealed is a sobering one. Table 10.10 and Figure 10.13 illustrate the increasing strain on water resources which Iraq must inevitably feel, if all the Turkish and Syrian projects were to be realized.

It will be noted that the minimum amount of water received by Iraq varies from 4,088 Mm^3 in Figure 10.14 and Table 10.11 to 4,473 Mm^3 in Figure 10.1 and Table 10.10. This difference stems from a more exact accounting for return flow in the former case. It also should be kept in mind that "natural flow" and actual river conditions seldom coincide. Moreover, year-to-year fluctuations, such as those discussed in Chapters 5, 6, and the previous section of this chapter, further complicate matters, especially if they coincide with reservoir filling or, conversely, include exceptionally large flood stages. Nevertheless, the general pattern of steadily impending crisis is clear.

11

The Years Ahead

Six years have passed since initial data collection for this study began. During that time a more complete picture has emerged from the first—often glowing, often overly optimistic—reports issued by Turkey and Syria regarding their respective plans for the Euphrates River. Nevertheless, the world at large has become increasingly aware that water, not oil, will be the key to peace and war in the Middle East. In November of 1988 scenarios regarding the future impact of water shortages throughout the region reached a frightening, if fanciful, level with an article in *U.S. News and World Report* (Chesnoff 1988). To quote briefly: "Nov. 12, 1993. War erupted throughout the Middle East today in a desperate struggle for dwindling water supplies. Iraqi forces, attempting to smash a Syrian blockade, launched massive attacks on the Euphrates River valley. Syria answered with missile attacks on Baghdad."

A Realistic Prognosis

The conclusions given at the end of the preceding chapter might seem to substantiate such a grim picture, but it should be remembered that the projections described in Chapter 10 are based on the assumption that all of the Turkish plans and most of the

Syrian plans will come to fruition on the tight schedule indicated by the charts, tables, and diagrams included in that chapter. Those projections indicate serious water shortages sometime shortly after the year 2000. (The scenario proposed in the *U.S. News and World Report* article advances that date by about a decade.)

Careful consideration of the data and time frames presented in the preceding chapters indicates that less land will be irrigated than first surmised and that many projects will come on line considerably further in the future than predicted. In fact, interviews conducted in the summer of 1988 with GAP engineers and administrators and other interested parties suggest that 2030 might be a more realistic date for the completion of GAP and that, even in Turkey, less land may receive water than was first intended. In the case of Syria, evidence is accumulating that the Euphrates projects there will be significantly reduced and are even now being cut back.

Other issues will inevitably affect the ultimate dimensions of GAP and the Syrian Euphrates. In Turkey, overall project management and coordination is, at the time of this writing, being articulated. Beyond the problems of finishing existing projects and beginning the remaining ones are complications relating to production and marketing decisions and the final disposal of whatever crops are grown. Inequalities of land ownership are also attracting attention, and the question of the Kurdish and Arabic speaking minorities within the region must be addressed. Even the initial authorship of the idea of GAP has become a matter of some interest, particularly in the Turkish press and in view of continuing political developments in Turkey.

While it is not the purpose of this study to consider the latter topics in detail (a detailed discussion of them can be found in a future volume in this series), it is necessary to try to refine, in terms of most recent developments, "the general pattern of steadily impending crisis" with which Chapter 10 ended. Because of the dynamic nature of the situation, even such "final" comments will need further clarification by the time these pages reach the reader's eyes.

The Turks, moreover, have long been known for their diplomatic acumen, and here too they are showing their awareness of the delicacy and importance of the impact of the projects they are undertaking on the Euphrates and the Tigris rivers. This chapter concludes with a brief consideration of Turkey's "Peace Pipeline" proposal, a bold and imaginative approach to resolving some, if not all, of the water problems of Southwest Asia. It is on this note of hope, intimately intertwined with GAP, that the study ends.

Status of GAP, September 1988

In order to attempt to predict how much land will ultimately be irrigated by GAP, as well as when various projects—river water depletions—will come on line, it is necessary to review the sub-projects which have been described and to list their status as indicated by press reports during the month of September 1988.[1]

All comments which follow refer to this time period unless otherwise noted.

• The Keban Dam, Hydroelectric Plant, and Reservoir. Although this project is not part of GAP, it is closely related to all downstream events on the Euphrates. The heavy rains of spring and summer 1988 filled the Keban Reservoir; hydroelectric production continues. No significant problems are apparent, though unverified comments regarding the need for maintenance and overhauling of the generators at the power plant persist.

• The Karakaya Dam, Hydroelectric Generating Plant, and Reservoir. The dam has been completed; the reservoir is full; five generating units were installed by early 1988 and the sixth and last came on line in November 1988.

• The Lower Euphrates Project, which includes the Ataturk Dam, the Urfa Tunnels, the Urfa Hydroelectric Generating Plant, associated canals and irrigation projects.

---

1. Reference is made particularly to a series of feature articles by Ismet Berkan which appeared in *Cumhuriyet* at that time. Other sources (*Ekspres, Huriyet, Turkish Daily News*) were consulted where appropriate. As will be seen, there is internal consistency between the analytical descriptions of this study and the aforementioned reports. The conclusions reached in this section are also substantiated by field observations made by Kolars during the summer of 1988.

Work is progressing on the Ataturk Dam at an uneven pace (see discussion in the section which follows). Estimates now indicate that the dam will be finished in 1991–1992.

The Urfa Tunnels are to provide 328 m³ (2 x 164 m³) of water by gravity flow to the Urfa-Harran area. Early estimates predicted that the tunnels would be completed sometime in 1988, but technical difficulties of an undefined nature, plus lack of funds, now indicate that the tunnels will be finished by Spring 1991 at the earliest. As of September 1988 the tunnels were 64 percent complete; that is, 17 km of the intended 26.4 km had been excavated and about 25 percent lined with concrete. Twelve access shafts were in operation, and another four were planned. Working, housing, and commissary conditions for the tunnel laborers were reported by the press to be below minimum standards. Payrolls were also reported delayed.

Concern has been voiced in the Turkish press that, because the tunnels might not be completed until after the main portion of the Ataturk Dam is completed (scheduled for the Summer of 1990), water would be impounded above the level of the tunnel entrances, thus complicating their completion. In any event, the tunnels and the dam are of nearly equal importance. Many of the optimistic prognostications of GAP's role in the Turkish economy and of its ability to pay for itself are based on accompanying irrigation projects coming into production. Thus the tunnels occupy an important position in GAP.

The Urfa Hydroelectric Generating Station is located at the exit of the Urfa Tunnels and is intended to generate energy for local use. Work on this was contracted in 1985 and is continuing.

Three major canals have been contracted for the use of the waters of the Urfa Tunnels—one contracted in 1980, one in 1985, and one in 1986. These are scheduled to be finished by 1991 in time to receive water from the completed tunnels. Gravity flow should provide irrigation for 142,000 ha. This figure differs slightly from those in official documents (Table 11.1), but the discrepancies are so slight that there is no apparent reason to revise the data given earlier in this study. The characteristics of a sub-unit found on the Harran Plain are given in Table 11.2. Other reports suggest that water will arrive in the Harran area in 1992.

Table 11.1. GAP Comparative Hectarages to Be Irrigated, 1980 and 1987
(in hectares)

Location	1980	1987
Urfa-Harran	157,000	142,000
Mardin-Ceylanpinar	140,000 (gravity flow)	150,000 (gravity flow)
	192,100 (pumped)*	178,000 (pumped)
Siverek-Hilvan/Bozova	164,300	
	55,300	
	219,600 (combined figures)	215,000 (combined figures)
Totals	708,700**	685,000

Source: 1980 data from GAP (1980); 1987 data from *Cumhuriyet* (1987b).

*These figures correspond to the Derik-Mardin area.
**For a complete listing for projects see Table 11.4.

At some unspecified later date contracts will be tendered for additional canals to provide water to the Mardin-Ceylanpinar area from the same tunnels. Approximately 140,000 ha will be served by gravity flow and another 192,100 by additional pumping.

Preliminary studies are underway regarding the Siverek-Hilvan and Bozova Projects, both of which will rely upon water pumped to elevations above the reservoir. Press reports show a combined total of 219,600 ha for these two schemes. No firm date is scheduled for beginning work on these projects.

● Euphrates Border Project: the Birecik and Karkamis Dams. Work was programmed to begin on these in 1987, but to date nothing has begun.

● Siric-Baziki Project. This project is in the "preliminary study" stage of development.

● Adiyaman-Kahta Project. This project is at the "master plan" stage of development.

● Goksu-Araban Project. Work on this project was reported to have begun in 1990.

● Gaziantep Project. For the Hancagiz Dam, construction

Table 11.2. A Sub-Unit in the Harran Plain Irrigation Area

Ahmet Aydeniz
Sanliurfa Irrigation
2d Division Construction
Celik Kalip Workshop
Unit 1

General Directorate of General Hydraulic Works
XVth Region—Sanliurfa Plain Irrigation

Engineer: Ahmet Aydeniz
Estimated Cost: 10,300,000,000 TL
Discount: 2.86%
Discounted contract cost: 10,005,420,000 TL
Area to be irrigated: 24,920 ha
Small manufactured channels—total length: 700,035 m
Mfctd channel production capacity: 600 m/day
Date of work completion: 30/10/90

Assorted types of small canal lines	402
Siphons	1,502
Canal separators (i.e. flow dividers)	54
Chutes	191
Canal end structures	402
Canal elbows	831

Source: Information translated verbatim from a signboard south of Urfa near Harran, May 1988.

Note: cost per ha = 401,501.6 TL; cost per meter of canal = 14,292.7 TL

costs according to the 1985 contract were calculated to be TL 30 billion. Money spent to September 1988 amounts to TL 59 million, less than 0.2 percent of the total cost. Work on pumping stations "possibly will begin next year." Rumor reports a similar situation for the Kayacik Dam.

The following comments refer to the Tigris River portions of GAP.

• Kralkizi and Dicle (Tigris) Dams. Work on these two dams is continuing. The irrigation canals will be built later.

● Batman Dam. Work on the dam is continuing. One diversion tunnel is completed, and work on the second is in progress. Construction on the main body of the dam will begin later.

● Batman-Silvan. More preliminary studies are being conducted.

● Garzan. Preliminary studies are continuing.

● Ilisu Dam. This project, which threatens to inundate the historic town of Hasankeyf, is still only in the planning stage. Work will begin on it "in the years ahead."

● Cizre. Work on this dam was scheduled to begin in 1989. Irrigation canals will be constructed in 1992.

A Revised Schedule for GAP

A revised schedule of dam construction, reservoir filling, and irrigation works is shown in Table 11.3. The material shown in this table reflects the original estimates for various projects as shown in Table 10.4, but the sequence and timing of events as shown is based upon more recent reports and upon an assessment of the financial and planning aspects of GAP as discussed in the section that follows. While all dates are conjectural, the extension of completion time(s) into the twenty-first century could result not only from technical difficulties, both real (such as problems with the Urfa Tunnels) and anticipated, but also from financial shortages (discussed in the following section) and the largely unanswered question of what crops can be grown which will find a ready and profitable market within or outside Turkey.

Informal talks with a number of concerned Turks indicate that the extension of completion dates from thirty to fifty years for the entire project is entirely reasonable. Comparison of Figure 10.14 with Table 11.3 and Figure 11.1 shows the anticipated depletion of approximately 10 billion m^3 of Euphrates water occurring not by 1990, but more likely sometime near the year 2010. In the same way, depletion of 12.5 billion m^3 may be delayed until nearly 2020, thirty years after the initially projected date. And again the absolute depletion of over 16 billion m^3 may occur sometime around 2040.

Table 11.3. GAP Projects and River Depletions—Revised Estimated Schedule

Year*	Project*	Estimated Total Depletion 1000s m³	Aggregate Total 1000s m³
1988	Keban Reservoir**	986	986
1988	Karakaya Reservoir**	435	1,421
1990	Small upstream projects	241	1,662
1991	Ataturk Reservoir	1,471	3,133
1992–1995	Harran Plain	1,688	4,821
	Misc. small projects	232	5,053
Post 1995			
2000	Mardin-Ceylanpinar (gravity flow)	1,606	6,659
	Misc. small projects	165	6,824
2005	Derik-Mardin-Ceylanpinar (pumped)	2,157	8,981
	Gaziantep (see also below)	294	9,275
2010	Siverek-Hilvan (pumped)	1,539	10,814
2015	Bozova	595	11,409
	Misc. small projects	215	11,624
2020	Karkamis	57	11,681
	Birecik	113	11,794
	Goksu	344	12,138
	Araban	247	12,385
2025	Adiyaman-Kahta	1,433	13,818
2030	Gaziantep (pumped from Birecik Res.)	547	14,365
2040	Siric-Baziki	1,567	15,941
	Misc. remaining projects	967	16,908

Source: Table 11.4. See text for comments on sequence and timing events.

*Matching of dates and projects conjectural except for 1988.
**In place, having been completed earlier.

A more immediate and predictable problem will accompany the irrigation of the Harran Plain and the Ceylanpinar area. Work in both of these projects is underway and, with the filling of the Ataturk Reservoir and the completion of the Urfa Tunnels, there will be an almost automatic demand that the water be used because it is available. This, in turn, will lead directly to increased return flow, first downstream on the Balikh River and thereafter in the Jagh Jagh drainage area, both in Syria. It has already been shown that such return flow must inevitably present problems of pollution to downstream users. Turkish engineers are aware of this.

Although no specific public discussion of the problem has come to the attention of the authors of this study, informal talks have suggested a number of solutions: impoundments to retain dirty water; bypass canals, either west to the Euphrates in Turkey or south to the Euphrates below Deir ez-Zor; injection below ground (though the problem of the Ras al-Ayn aquifer must be confronted, if this were to be done); or possibly free flow downstream—certainly the least acceptable, laissez faire approach.[2]

Whatever the outcome to the above questions, their general effect must be to slow the pace of development of GAP in its totality. Before turning to similar problems in Syria, a brief review of financial and other matters in Turkey is in order, although a complete financial analysis is not the intent of the present study.

Financial and Managerial Aspects of GAP

Development of the Euphrates River began modestly. The idea of GAP began to acquire a resonance of its own with work on the Karakaya Dam and was fully articulated for public appraisal and approval with the prospectus issued by the DSI (GAP 1980). In the years that followed, more and more attention has been directed to the use of the irrigation water to be made available. Glowing

---

2. It should be noted that routing the return flow west would create problems with the waters of Lake Assad, while routing the return flow to below Deir es-Zor would simply pass such problems along to the Iraqis.

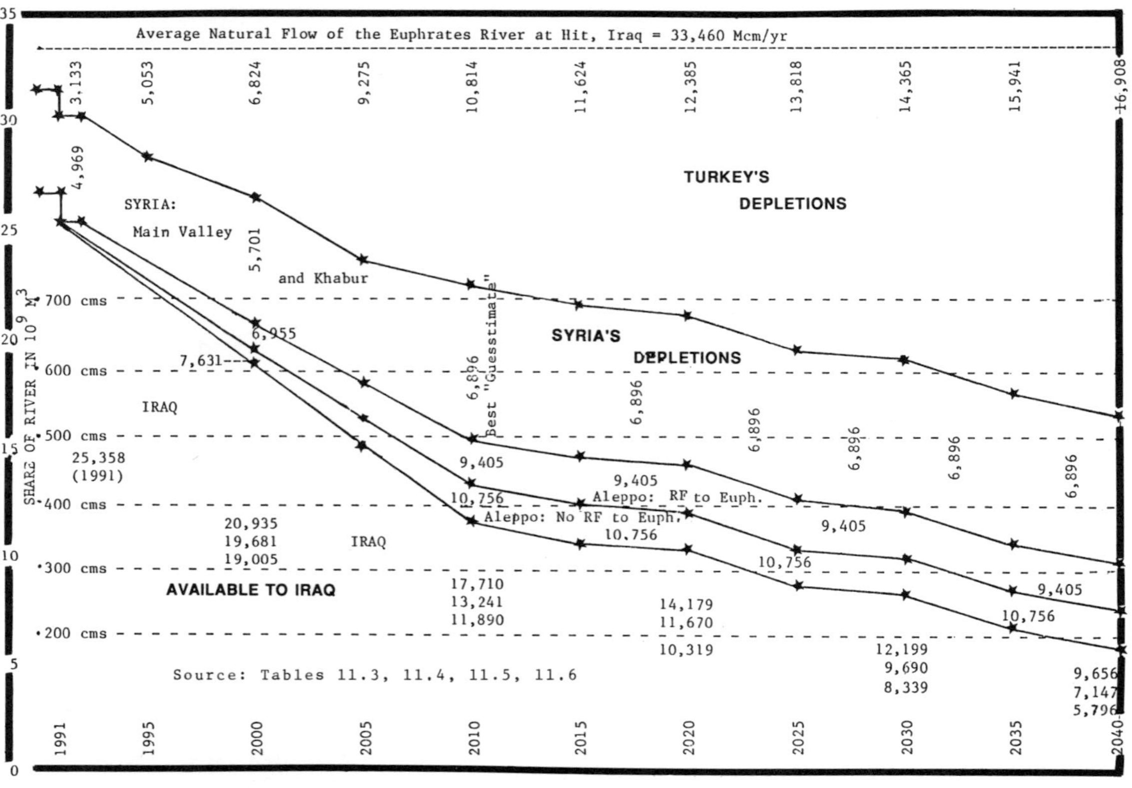

Figure 11.1. Projected sequential depletion of the Euphrates River, 1990–2040.

This figure summarizes the analysis described by the foregoing text. Each value has been carefully derived. The timing of these events is more speculative than the data themselves and represents, at best, informed opinion and not fact. Nevertheless, the combination of data, analysis, and opinion presented here gives a unique view of impact of the developments proposed and underway along the Euphrates River in both Turkey and Syria.

This representation may be considered predictive in two ways. First, the increasing depletion of the Euphrates' waters can be read from left to right. Second, the intersections of the m^3/s measures (shown by dashed lines labeled "cms"), with lines representing removals, indicate the year in which certain levels of flow may be reached. If 500 m^3/s entering either Syria or Iraq from its upstream neighbor is taken as the minimum flow acceptable to either country, it can be seen that under the circumstances postulated here, Syria should not be shorted by Turkey. (Bear in mind, however, that the pattern of flow—either by the mainstream or via the Urfa tunnels—will have much to do with whether or not conditions remain felicitous for Syria.) On the other hand, Iraq may feel the pinch as early as 2005 if the Aleppo project without RF were to occur—or around 2010 if only the main valley and Khabur projects are realized.

Values relating to the Aleppo project need further explanation. As mentioned earlier, hectarage between 180,000 and 212,000 has recently been proposed for the area north and south of Aleppo. (This new development has not been considered in any detail in the preceding chapters but is included here for the sake of completeness.) Water for these fields would be taken from Lake Assad. For simplicity, a round figure of 200,000 ha has been used to compute depletion and RF from this project if fully implemented. Depletion of 12,545 m^3/ha and 6,755 m^3/ha RF are based upon values computed for similar areas nearby which are used elsewhere in this text. Such removals and returns are assumed to begin about 1991 and to increase steadily until the full 200,000 ha are under irrigation in 2010. Two lines are used to depict such a situation. The one showing depletions amounting to 9,405 Mm^3/yr represents what would happen if RF from these fields reaches the main stream of the Euphrates—thus restoring some of the water removed. That this may happen is uncertain. The area around Aleppo is essentially a basin of interior drainage which might trap drainage preventing its return to the main stream. In this case, the RF lost to evapotranspiration or seepage would be a net loss to the Euphrates system, which along with regular evapotranspiration losses would amount to about 10,756 Mm^3/yr as shown by the lowest of the three lines depicting Syrian removals. It should be further noted that Syrian removals are expected to stabilize about 2010 with no increase or decrease thereafter. (This overlooks possible future loses of land resulting from poor drainage and soil salinization.)

It is unlikely that the worst-case scenario shown for 2040 will ever be reached. But if it were to be realized, Iraq might expect less than 200 m^3/s to enter across its border from Syria.

Needless to say, the reader must keep in mind the highly conjectural nature of all these speculations.

projections of its impact on agricultural production in Turkey, in addition to the undeniable utility of the electricity produced by the new and anticipated dams, made credit for its conception a sought-after political prize, although in recent months much criticism has also been leveled at the cost of GAP.

According to the Turkish press, claims have been made by both Turgut Ozal and Suleyman Demirel that each deserved the credit for GAP, Demirel for activity during his time as Prime Minister and Ozal both as an electrical engineer and planner for the DSI and as Prime Minister. Other parties, such as the Democratic Populist Party (SHP), have not taken sides and refer to the project simply as "government property." Human interest was added to all this when the former mayor of Urfa, Cemil Hacikamiloglu, recently claimed authorship by citing ideas which he had presented in the early 1960s to both Celal Bayar and Suleyman Demirel for the use of Euphrates waters near Urfa.

More serious issues concerning GAP began to reach the public in late 1987 and 1988. In March 1988, *Cumhuriyet* reported that Ata Insaat (the consortium building the Ataturk Dam) had received only TL 80 billion but anticipated doing at least TL 200 billion worth of work in 1988. The Akpinar Group (responsible for the Urfa Tunnels) was reported to have received approximately TL 10 billion and to be nearly three years behind schedule. In June 1988, there were 8,200 workers filling two shifts on the Ataturk Dam, but, in July, 7,000 of these were laid off for 20 days because of unpaid receipts of TL 150 billion owed Ata Insaat by the government. Such payments reportedly were originally scheduled for the previous January.

At a briefing given to President Evren during his visit to GAP in June 1988, Assistant Permanent Undersecretary of the State Planning Organization, Dogan Yorukhan, recalled the original estimate for GAP's completion was 1994. He then pointed out that, to achieve this deadline, TL 64.6 trillion (1×10^{12}) would need to be invested, including a current need for TL 9.4 trillion. Meanwhile, in April of 1988, additional money was allocated to the completion of 13 dams in Turkey from government "Public Partnership" funds. Of the TL 74 billion (1×10^{9}) thus designated, TL 18.9 billion went to the Ataturk Dam and TL 2.5 billion to the

Karakaya project. Ministry of Housing and Public Works Director Bulent Gultekin announced that, by the end of 1988, the government's investment in GAP would reach TL 230 billion.[3]) However, as late as 23 August of the same year, it was reported that the government owed contractors working on GAP approximately TL 125 billion.

In 1983 GAP contracts for the Ataturk Dam were estimated at TL 102 billion and for the Urfa Tunnels (at 1981 prices) TL 27 billion, but, by August 1988, work was reportedly estimated to be costing TL 1 billion per day. As Ismet Berkan pointed out in *Cumhuriyet* in September 1988, the biggest problem facing GAP is the financial one. To paraphrase his words, no more than one-fourth of the Lower Euphrates Project is finished or under construction. This is not to mention other projects, such as the Karkamis and Birecik Dams and the many projects scheduled for the Tigris. Some of these have already been started (see summary given earlier in this chapter), while others have not yet even been contracted.

Nevertheless, there is little doubt that, with so much at stake both financially and politically, as much as possible of GAP will be completed as money becomes available. It is also reasonably safe to say that the Ataturk Dam and reservoir and the Urfa Tunnels and the gravity flow projects associated with them will come on line in the near future. On the other hand, the many additional projects described and analyzed in this study may be delayed for years or perhaps decades, not through technical difficulties but for the reasons touched upon here.

Another set of problems must also be considered in passing. These relate to the articulation of the projects, their coordination, and their eventual realization with products for the domestic and world markets. Even the timing of technological events has evoked concern, as in the case of the completion of the Urfa Tunnels and the filling of the Ataturk Reservoir. A similar situation concerns the height to which the Ataturk Reservoir might be filled. During President Evren's visit to GAP, the General Director

3. By December 1988, the exchange rate was approximately $1.00 = TL 1,700.

of DSI observed that, if it were filled an additional 30 m, its useful life could be extended for fifty years. However, such a revision cannot be lightly undertaken because of the difficulty in altering original plans for the dam as well as the flooding of additional, valuable valley fields.

In response to these and other issues concerning project management, President Evren suggested that a special Ministry for GAP be established within the government. Very little has come of this suggestion to date. This may be due in part to the national referendum, actually a vote of confidence, held in the Autumn of 1988, but it is more likely that other measures described below were felt to meet the need for coordination and further study of the many problems generated by GAP. In any event, President Ozal has maintained control of the government at the time of writing and is continuing work on GAP as steadily as possible. (An indication of his commitment is that his official airplane is named "GAP.")

The government response to calls for better management of GAP was to create a Project Management Unit (PMU) under the responsibility of the State Planning Organization in late Fall of 1986. The Undersecretariat for Research and Project Promotion is responsible for the PMU's coordination and supervision. The PMU is to seek out international consulting firms with suitable experience related to GAP and its problems. The PMU also is to identify and prepare the terms of reference of the studies to be conducted by these consulting firms and to supervise the work they carry out. A main office of the PMU, located in Sanliurfa, and a liaison office in Ankara design and plan studies of the following:

1. Review of the "existing social and economic structure and resources of the region" and of the problems and potentials thereof
2. Alternative modes for industrial development offered by the region, including agriculture and trade
3. Identification of future policies for rural and urban development of the region
4. Investment opportunities and the magnitude needed for social and economic development of the region

5. Identification of regional development strategies and financing strategies for planning, programming, and budgeting
6. Planning of investments for rural areas
7. Development of soil, water, manpower, and other resources in rural areas, as well as social, economic, and physical ones
8. Securing positive urban development through the control and direction of investments within GAP

The PMU's first task was to call for bids for a GAP Master Plan contract. The work was eventually secured by the Japanese firm of Nippon Koei, which has subsequently established offices in Urfa. They were specifically enjoined to develop an Agricultural Production Design aimed at maximizing the production of both export and domestic crops. (A penultimate draft of this design was completed by the summer of 1988 and the final report submitted in April 1989.)[4]

Studies have been underway to establish an Economic Development Agency for the promotion of investments in the region. There are also plans for an employment agency, both to create jobs and to find workers for jobs as they occur.

Other works which are underway include: an Irrigated Agriculture Research and Education Group to be headquartered in Urfa; the improvement of the airport at Urfa; and a cooperative effort between the Ceylanpinar Agricultural Station, the various offices in Urfa, and the Agricultural Division of Cukurova University which will carry out experiments to determine the best crops and methods of planting in the areas to be irrigated in the near future.

In addition to the many official actions, some of which are described above, independent, private initiative is at work within the region. Many large landholders within the area have already invested in gasoline pumps and plastic irrigation pipes served by wells drilled on the Harran Plain south of Urfa. Several thousand

4. Nipon Koei Co. Ltd. and Yukesel Proje A. S. Joint Venture, vols I-IV (Tokyo and Ankara), GAP, The Southeastern Anatolia Project Master Plan Study, Final Master Plan Report (April 1989).

hectares of land were being irrigated in this manner in the summer of 1988, and shops selling similar equipment were to be seen on all sides in Urfa. Despite avowals by some of the landlords that their tenant farmers could not be taught the necessary techniques, it was evident that the new technology was diffusing rapidly throughout the region.

This, in turn, raises questions of the possible diminution or exhaustion of the aquifer underlying the area, which in turn feeds Syrian springs to the south. As mentioned elsewhere, return flow from these fields also may begin to influence water quality in Syria. Another item of concern noted during a field reconnaissance in the summer of 1988 was the profligate use of water in unlined ditches being applied to fields by-guess-and-by-gosh, rather than under the supervision of trained irrigation experts.

In summary, the many elements described above indicate a situation in a state of rapid development and flux, one wherein little more than impressions can be had until a more stable second phase of GAP is established. As stated above, many parts of GAP will come on line, some slowed, some expedited, by current conditions. All of these, in turn, will impact upon Syrian plans and activities downstream on the Euphrates and its tributaries.

It is now time to re-examine the Syrian situation and to predict what is to come.

Future of Syrian Euphrates Development

By the Summer of 1986 a pattern of development along the Syrian Euphrates had begun to emerge. Original plans full of high hopes to irrigate 640,000 ha of land were revised downward drastically. Almost all of these downward revisions were the result of unexpected problems relating to gypsiferous soils, which dissolved upon contact with water leaking from new canals, thus disrupting the entire system as it was being put in place. Recent estimates by the Syrian government indicate that only from 240,000 to perhaps 260,000 ha of land will be irrigated when the projects have reached completion.[5]

The significance of these figures is diminished when it is realized that 150,000 ha were irrigated by private pumping prior to the beginning of the government program. Moreover, although 64,500 ha had been lost to the floodwaters of Lake Assad, another 31,000 ha of irrigated land had been disrupted by land reclamation works, and another 7,495 ha of dry-farming land had also been flooded (Meliczek 1987, 110).

Table 11.4 summarizes the ongoing status of the projects attempted and realized to date by Syria. Columns 2, 3, 4 of Part I, "Proposed, but Unrealized," represent estimates of land studied "generally" and "in detail" which were designated in 1977. These exclude the Resafe (124,300 ha), Mayadin (100,000 ha), and Lower Khabur (370,000 ha) units which have either been abandoned (Resafe and Mayadin) or suspended from further development. This shows a major cutback from the original 640,000 ha.

Columns 5, 6, 7 show the areas and amounts of land being carefully studied, worked upon, or actually in irrigated production, circa 1982–1983. This new total, while smaller than the previous one, includes 54,000 ha scheduled for production on the Upper Khabur. This latter area does not and will not draw water directly from Lake Assad, but rather will depend upon waters of the Khabur/Habur originating in Turkey. While all of this land may not come into eventual production, the total for the main valley of the Euphrates approaches that cited by Metral of 269,000 ha versus the government's estimate of 240,000 ha.

Columns 8, 9, 10 of Part III show the amounts of land actually in production through irrigation in 1982–1983. This amount is consistent with the slightly larger area (64,500 ha) reported in 1986 by Meliczek. It should be kept in mind that significant amounts of privately irrigated land were still being farmed in 1982. Estimates of these are shown in Table 11.5.

5. Meliczek (1987, 129) reports a figure of 240,000 ha; close accounting based on Metral's (1987) data (p. 119) places this amount at 259,300. No definite way exists at present to resolve this difference.

Table 11.4. Estimated Project Water Depletion and Return Flow: The Syrian Euphrates and Tributaries
(area in hectares, water flow and losses in Mm³)

Project Name	I. Proposed in 1977			II. Proposed/Actual 1982–83			III. In Production 1982–83		
	Irrig Area*	Return Flow	Water Loss	Irrig Area	Return Flow	Water Loss	Irrig Area	Return Flow	Water Loss
Sajur	?	?	?	?	?	?	?	?	?
Tishreen Dam (reservoir evap)	—	—	157.5	—	—	—	—	—	—
Aleppo diversion	—	—	80.2	—	—	80.2	—	—	80.2
Lake Assad (reservoir evap)	—	—	1,570.0	—	—	1,570.0	—	—	1,570.0
Baath Dam (reservoir evap)	?	?	?	?	?	?	?	?	?
Maskanah	116,000 (12,545)	783.4	1,455	74,300	591.2	932.1	24,300	163.9	304.8
Balikh #1	191,600 (10,754)	1,109.2	2,060.5	106,000	613.7	1,140	17,000	98.4	182.8

Middle Euphrates	29,700 (16,835)	280.5	500.0	29,000	273.9	488.2	12,000	108.8	202.0
Lower Euphrates	123,000 (16,835)	1,115.2	2,070.7	50,000	453.3	841.8	none	none	none
Upper Khabur	—	—	—	54,000 (12,017)	349.4	648.9	none	none	none
TOTAL	460,300	3,288.3	7,893.9	313,300	2,180.5	5,701.2	53,000	371.1	2,339.8

Source: Metral (1987, 119, Table 1).

*Depletion rate in m^3/ha/yr is shown in parentheses.

Table 11.5. Syrian Project and Private Water Depletion and Return Flow,* ca 1982–83

	High Private Estimate	Low Private Estimate
Depletion rate	13,378 m^3/ha**	13,378 m^3/ha**
Total area	196,000 ha	176,500 ha
Total depletion per year	1 Mm^3	2,361.2 Mm^3
Project area	53,300 ha	53,000 ha
Total project depletion per year	2,339.8 Mm^3***	2,339.8 Mm^3***
Total depletion private & public use	4,961.9	4,701.0 Mm^3

*Project depletion based on "in production" figures, col 10, Table 11.5
**Average of three depletion rates; see col. 2, Table 11.4
***Includes Lake Assad and Aleppo diversion.

While it may be assumed that such private enterprises will eventually be incorporated into state run projects, that must remain for the future. Table 11.6 attempts to estimate what the end result will be. It is somewhat facetiously called a "Best Guesstimate," because of the many political, social, and financial factors which may intervene, in addition to even more unforeseen physical constraints upon these projects.

Nevertheless, this table presents what is considered to be a most likely future situation. The hectarage shown for the main valley (259,300 ha) is thought to be consistent with the 240,000 ha figure cited earlier, but derived with a finer attention to detail. To this is added the 137,900 ha planned for the Upper Khabur. While these latter areas will depend to a great degree upon what happens upstream in Turkey, if all goes optimally there, the land and water is available in the High Jezirah to sustain this amount of new irrigated farmland. Thus, while the total may err on the side of optimism (do not forget that the Upper Khabur portion of this grand total was not included in any estimate of hectarage for the main valley), it is this amount of irrigated land which has been

Table 11.6. Public Project and Private Water Depletion and Return Flows Upon Completion of All Projects
(best 'guesstimate')*

Project Name	Land Irrigated in ha	Return Flow in Mm^3	Total Water Loss in Mm^3
Sajur	?	?	?
Tishreen Reservoir	—	— evaporation from surface	157.2
Aleppo Diversion	—	—	80.2**
Tabqa Dam Reservoir	—	— evaporation from surface	1,570.0
Baath Dam Reservoir		not included	
Maskanah	74,300	501.2	932.0
Balikh	106,000 (10,754)	613.7	1,140.0
Middle Euphrates	29,000 (16,835)	262.9	488.2
Lower Euphrates	50,000 (16,835)	453.3	841.8
Sub-total	259,300	1,831.1	5,209.8
Upper Khabur	42,000 (11,489)	260.0	483.0
(See Table 11.4)	49,450 (12,545)	648.0	1,203.0
	46,450 (12,545)	combined	combined
Total	397,200 ha	2,739.1 Mm^3	6,895.8++

Source: Meliczek (1987, 119); Table 11.4, this study.

*All private shares are assumed to be integrated for this estimate.
**Rapid urbanization of the Aleppo area will undoubtedly increase this figure. Also, Meliczek lists but does not discuss 212,000 ha in the Aleppo area that may possibly be irrigated all or in part. Such a development would, naturally, dramatically increase the figures shown herein.

used to estimate ultimate water depletion resulting from Syrian activities sometime in the future.

The various depletions which might result from the several total hectarages discussed are shown in Table 11.7. Those above the "Best Guesstimate" in column 1 are unattainable. Those figures appearing below it in the column represent possible interim quantities or, perhaps, stopping points, if all does not go well. Given these estimates, it is now possible to return to Figure 11.1 showing the total estimated amounts of water to be removed from the Euphrates River in Turkey and Syria before entering Iraq. As mentioned previously, the estimated amounts shown for Turkey appear to be attainable, although the scheduling of such depletion events will be extended into the twenty-first century.

Perhaps the safest estimate of realizable projects for Turkey, all else being equal, would be the conditions shown occurring between 2015 and 2020—that is, depletions amounting to 12,385 Mm^3/yr. By the same token, the "Best Guesstimate" removals (6,896 Mm^3/yr) for Syria may not occur until the same time, although in this figure they are shown as completed by 2010. The extreme depletions shown in the 2000 to 2000+ column of Figure 10.1 will certainly not happen as far as Syria is concerned (12,079 Mm^3/yr), nor should such removals by Turkey (16,909 Mm^3/yr) be a cause for present concern. Thus the total figure which may eventually reach Iraq in 2020 (14,179 Mm^3/yr) is close to the amount asked for by that country (449 m^3/s cf. 500 m^3/s).[6]

By way of comparison, in the year 2000 according to this

6. A further caveat must be made at this point. Dr. Abduh Qasim, in an interview mentioned earlier (Chapter 8), has spoken of an additional 180,000 ha of land to be irrigated in the northern and southern Aleppo region. Meliczek (Table 11.6) lists, but does not comment on, a similar figure of 212,000 ha. While little else has appeared regarding these hectarages, Patrick Seale (1988, 445) states, "The Sixth Five-Year Plan, 1986–9 . . . diverted Euphrates water westward to the fertile plains north and south of Aleppo." If this has happened or happens, approximately 2,509 Mm^3/yr depletion of the river should be added to the amounts spoken of here. This estimate is based upon similar depletion values for the Aleppo-Maskanah sub-project discussed earlier (see Table 10.3, Part 2). Return flow—that is, additionally polluted water, as opposed to undiverted river flow—would also be increased by 1,351 Mm^3/yr. However, such return flow might be trapped topographically near Aleppo, thus raising this deficit increment to 3,860 Mm^3/yr (see note to Figure 11.1).

Table 11.7. Water Use on the Syrian Euphrates—Possible Water Depletions and Return Flows From Various Estimates

Estimate	Hectares Irrigated	Return Flow in Mm^3	Total Water Loss in Mm^3
Highest Planned Estimate (Table 10.4)	752,900	5,530	12,079
Proposed but Unrealized (Table 11.4 No. 1)	460,300	3,288	7,894
Best "Guesstimate" (Table 11.6)	397,200	2,739	6,896
Underway and Irrigated (Table 11.4 No. II)	313,300	2,181	5,701
Irrigated: 1982-83 High estimate, both public and private (Table 11.4 No. III)	249,300	—	4,969
As above: Low Estimate	229,800	—	4,701

Source: Tables 10.3, 11.4, 11.6 of this study.

revised schedule (see Table 11.3), Turkish depletions may amount to 6,824 Mm^3/yr and Syrian depletions to approximately 5,701 Mm^3/yr, allowing an estimated 20,935 Mm^3/yr (664 m^3/s) to enter Iraq. If the new Aleppo diversion mentioned in note 7 has come on line by that time, this amount could be reduced to 19,005 Mm^3/ yr or 603 m^3/s).

Whatever happens, the quality of such water can be seriously altered by contaminated return flows. Also it must not be forgotten that a significant portion of such flow will reach Iraq via the Urfa Tunnels and the Khabur and Jagh Jagh rivers, a fact that may present special problems for Syria.

Much of what has been said above relates to elusive and nearly intangible events and attitudes, and the equation may swing in either direction. That is, on the one hand, profligate use of water with little attention to conservation or quality will exacerbate the international situation; so, too, might a run of very dry years in the catchment area. On the other hand, the failure to attain large areas of irrigated land in either upstream country would also increase downstream flow; so, too, would runs of more humid years.

The salient fact remains that the use of the Euphrates and Tigris rivers by Turkey, Syria, and Iraq presents questions, problems, and opportunities which can only become more pressing in the years ahead. Time exists for the diplomatic negotiations needed to attain some balanced use of these vital resources, but that time is growing perilously short. Countries other than Turkey, Syria, and Iraq may also become involved.

The section which follows explores one way in which this might happen.

Turkey and the Middle East *Pax Aquarum*

A new paradigm is developing in southwest Asia. In the past, there were "have" nations with ample petroleum resources and "have not" nations, which were petroleum barren or petroleum poor. The nations of the Peninsula, as well as Iran and Iraq, were essentially the "haves", while Turkey, Syria,[7] Lebanon, Jordan, and Israel were "have nots." But times change, and new circumstances present a different paradigm in which food production and security based on ample supplies of water are beginning to weigh as heavily as petroleum profits in the international scales. Population pressure in southwest Asia is forcing to the fore the fact that Turkey is the only viable source of surplus water in the region and that on-going development of Turkish and Syrian hydrologic resources has already made this a critical issue for diplomacy.

Proof of the above is that, in the Middle East (excluding the Maghreb), 50 percent of the population depends upon surface waters which cross an international boundary from some other coun-

7. Recent discoveries of petroleum and natural gas near Deir es-Zor may change Syria's position vis-a-vis this issue.

try, 67 percent of the Arabic-speaking world depends upon water from non-Arab sources, while another 24 percent are dependent on springs and deep wells. Specifically the population of fourteen nations (including the UAE as one unit) in southwest Asia in 1983 totalled 145.7 million. By the year 2000, they are expected to increase (by conservative estimate) to 234 million, about 61 percent more (Table 11.8) (World Bank 1985). (It should be noted that Sudan and Egypt, while included in the accompanying illustrations, have been omitted from this discussion, because the Nile basin countries represent a separate set of conditions and problems.)

Figure 11.2, which shows the availability of water throughout the region, expands upon the above idea. Attention is called to the group of countries with no surface water. With the exception of Iraq and Iran, these are the "have" nations of the original paradigm. To their right on the figure is a second group, not actually part of this discussion, which may be considered "dependent" nations, for, although they are amply supplied with water, that water comes from across international borders. Jordan also falls within this group, for its two streams, the Yarmouk and the Jordan, are shared with neighbors. Next are found Israel, Syria, and Iraq, which have some supplies directly under their control—because of orographic rainfall in the hills and mountains of the Levant and the Zagros and Elburz Mountains. However, major proportions of their supplies flow from outside their political boundaries. Lebanon and Iran may be considered independent as far as water is concerned. Properly managed, those countries have sufficient water to meet their projected future needs.

At the far end of the graph is Turkey with an abundant surplus of water, all from catchments within its own borders. (There are three exceptions to this last statement, namely, the Orontes [Asi], the Coruh, and the Meric, but they are of relatively slight importance.) Map 11.1 shows the occurrence of surplus water throughout the region. From this distribution, the comments made above regarding self-sufficiency should be apparent.

A note on sources of water other than international rivers is in order. In Saudi Arabia (Royal Embassy of Saudi Arabia 1987; *MEED* 1984c; *MEED* 1985c; *MEED* 1988a; and *MEED* 1988b) underground aquifers provide at present spectacular results and the

Table 11.8. Population Growth in Middle Eastern and Northeast African Countries, 1983–2000 (estimated)

Country	Population 1983 1×10^6	Est. Population 2000 1×10^6	% Change
UAE	1.2	2.0	67
Kuwait	1.7	3.0	76
Saudi Arabia	10.4	19.1	83
Oman	1.1	2.0	82
Yemen AR	7.6	12.0	58
Yemen PDR	2.0	3.0	50
(Libya)	(3.4)	(7.0)	(106)
(Egypt)	(45.2)	(63.0)	(39)
(Sudan)	(20.8)	(33.0)	(59)
Jordan	3.2	6.0	87
Israel	4.1	5.0	22
Syria	9.6	17.0	77
Iraq	14.7	26.0	77
Lebanon	2.6	3.0	15
Iran	42.5	71.0	67
Turkey	45.0	65.0	44
Total (less L/E/S)	145.7	234.0	60.6
Total (including L/E/S)	215.1	337.0	56.7

Source: World Bank (1985).

promise of such in Libya with its Great Manmade River Project (*Wall Street Journal* 1985; Ahmad 1981; and Pearce 1984). This cannot and will not last. Just as the United States is on the verge of exhausting one of the world's largest aquifers, the Oglalla sandstone of the Great Plains (Reisner 1986, 452–472), so too are there signs that water tables are dropping rapidly wherever they are being tapped throughout the Middle East. The Azraq Oasis in Jordan is a minor but significant example of this already having happened (*The Middle East* 1987). Such underground supplies are not a solution to the problem.

By the same token, desalination of sea water would have to be accomplished at a fraction of current costs to justify its use for

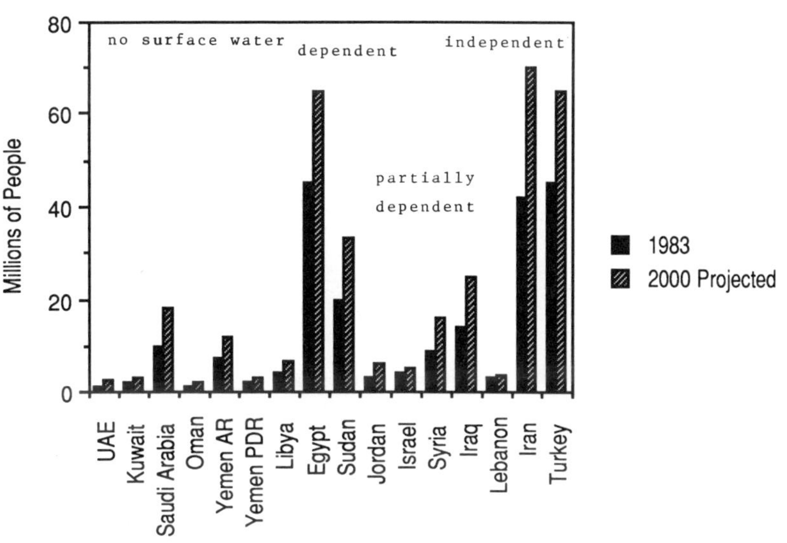

Figure 11.2. Population in SW Asia and NE Africa, 1983–2000 (estimated). Source: World Bank (1985).

agriculture. This works well for domestic purposes, particularly where ample supplies of energy exist as in the Emirates, Kuwait, and Saudi Arabia, but desalinized water for crop production, such as wheat and barley, requires ten to twelve times the quantities used by cities and industry. Moreover, seawater is desalinized at sea level, but it is used on fields at much higher elevations. In the case of Saudi Arabia, many of the agricultural areas now in place and producing are 200 to 500 m in elevation. Pumping costs can become prohibitive under these conditions.

In view of the above comments, a central development in this complex picture is GAP. As has been shown in the previous chapters, if every project listed by Turkey and scheduled for completion within the next thirty years were to come through, the impact upon the flow of the Euphrates would be dramatic. In turn, as Syria's plans for irrigation and hydropower come on line, additional depletions will take place from the Euphrates. This means that GAP will continue to put increasing strain upon the Euphrates and Tigris rivers, the two largest sources of water in southwest Asia.

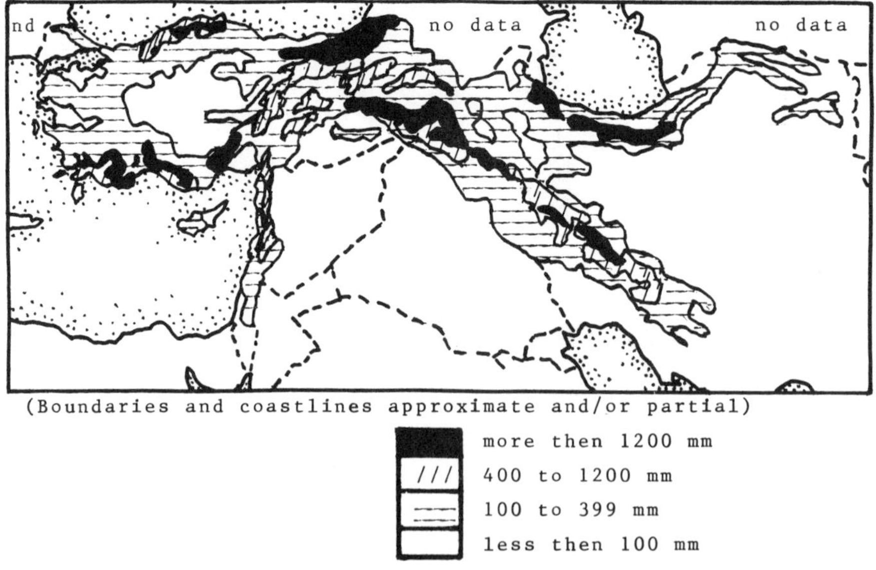

Map 11.1. Water surplus areas in the Middle East. Source: Thornthwaite et al. (1958).

Given increasing populations, lowering water tables, and GAP, what can be done? What is being done?

Negotiations between riparian users of the Euphrates have, in the past, encountered difficulties. At first, as a result of old disagreements, Turkey and Syria had little to say to each other. That situation has changed significantly since 1987, as Syrian officials have become increasingly aware of the implications which GAP holds for their country. In July 1987, Prime Minister Ozal visited Damascus and, upon his return, announced the signing of a protocol with Syria guaranteeing a minimum flow of the Euphrates of 500 m³/s across the border at Karkamis into Syria (*Cumhuriyet* 1987a).

This amounts to nearly 16 billion m³/yr, well within the range demanded by Syria in the earlier negotiations. On the other hand, this is the same amount demanded by Iraq in 1967 (Naff and Matson 1984). Syrian use will fall somewhere between 5 and 7 billion m^3 annually. If the larger figure is approached, then Iraq will be left in short supply. Thus a further solution to the problem must be found.

The possibility of success in these matters seemed for a while to be increasing. This was indicated by Syrian Ambassador to Turkey Abdul-Aziz al-Rifal's 1986 statement to the Turkish press, "Two neighbors do not struggle for water," and Iraqi Ambassador to Turkey Tariq Abdul-Jabar's statement in the same article, "The waters of the Euphrates and Tigris rivers will bring us a little closer together" (*Milliyet* 1986). A tripartite meeting of Syrian, Iraqi, and Turkish ministers was held in late November to discuss the problem of sharing the Euphrates's water (*TDN* 1988a; *TDN* 1988b; *TDN* 1988c; and *TDN* 1988d).

However, the *Turkish Daily News* and *Al-Hayat* (29 November, 1988) reported that, at this meeting, little was accomplished beyond the ministerial assumption of direction over a previously appointed technical committee intended to review the utilization of the waters shared by the three countries. By March 1989, the committee was to determine flow volumes—presumably with regard to both natural flow and the amounts to be removed by each riparian user. Turkish Public Works and Housing Minister Safa Giray was reported to have said that the committee's formation indicated Turkey's willingness to cooperate in the management of technical work on the waters of the Euphrates River.

The presence at this meeting of Iraqi Agriculture and Irrigation Minister Karim Hasan Rida and Syrian Minister of Irrigation Abdul-Rahman al-Madani signified the importance placed upon such talks and reflected Turkish diplomatic ability as well. Less clear was the alleged demand by Syria that the original 500 m^3/s—noted above—should be doubled. (This latter item may be the result of inaccurate reporting, for no other reference to it has been found).

Of much greater interest is *Al-Hayat*'s reference to a meeting between Turkish Minister of State Mehmet Yazar and Syrian Deputy Prime Minister for Economic Affairs Salim Yasin in Ankara early in November 1989. It was reported that an accord was signed at that time calling for the joint construction of a dam on the Euphrates River on the Syrian side of their mutual border. At first glance, this might appear to be the Tishrin Dam (mentioned previously in this study), construction on which is scheduled to begin in 1989 and to be completed in 1992. (This dam corresponds to the Yusuf Pasha Dam, which was proposed for a nearby site, but never built.) However, the article continues that, to date, "no feasibility studies or surveys have been carried out." This excludes the Tishrin Dam from consideration and indicates that it is possible that a new dam is being proposed.

The question then becomes: where might such construction take place? Any impoundment of water in Syria upstream from the Tishrin site would very likely flood Turkish territory, and, since the Karkamis Dam is planned in Turkey close to the Syrian border, one possible conclusion is that the Turks may have decided, for both diplomatic and austerity reasons, to scrap the Karkamis program in favor of something south of their border. This would have little effect upon GAP's total planned productivity or upon river flow, for the Karkamis was/is to be solely a small hydroelectric facility with estimated water depletion resulting from evaporation from its reservoir amounting to about 57 $Mm^3/$yr (see Table 10.4.) Moreover, similar or greater losses might occur from a more southerly reservoir's being built in Syria. Thus the overall change in effect upon the Euphrates—and upon the comments made in this analysis—would be negligible.

According to *Al-Hayat*, Turkish Minister of Finance Ahmet Alptemocin, in a speech before the National Assembly in late October of 1988, announced that major expenditures were to be "minimized," except in health services, fuel oil, and fire prevention, and only essential investments in infrastructure and defense would remain untouched. The Ataturk Dam was one such program to be spared with its completion date remaining 1990–1991. In the same vein, an unspecified State Planning Office spokesman announced that investment in major infrastructural projects which have yet to start will either be much delayed or set aside indefinitely. Even projects relating to GAP which are underway will be rescheduled with later completion dates, the Ataturk Dam and Urfa Tunnels being exceptions to this decision.

Technical exchanges continued during the months that followed. Then in late 1989 Turkey announced plans to begin filling the reservoir behind the nearly completed Ataturk Dam. According to Turkish reports additional water was released downstream prior to 15 January 1990 when the gates on the dam were closed for 30 days. At that point the Euphrates was reduced to a trickle, much to the consternation of Iraq and Syria. Those two countries formed a brief and uneasy detente in order to attend a tripartite ministerial-level meeting in Ankara regarding the flow of the river. According to the Turkish press, the Syrian minister maintained a low profile during the talks while the Iraqis were emphatic that at least 500 m^3/s should pass into Iraq from Syria. The Turks responded that this was a technical matter to be worked out by experts. The Iraqis in turn insisted that the matter was political and that not only the water issue but other matters prevailing between the three nations should be discussed at the same time.

It was at this point that the meetings stalled. Before further negotiations could take place, the invasion of Kuwait occurred and the issue of the Euphrates and its waters was eclipsed. At the time of this writing, there is little more to report on the matter. On the other hand, the Turks appear to be sensitive to the issues involved and have repeatedly said that they do not intend to use the river as a political or military tool against their neighbors to the south. In fact, President Ozal has long been the proponent of a possible solution to the water question through his "Peace Pipeline" proposal.

Early in 1987 then Prime Minister Ozal suggested an answer to the escalating water shortages of countries to Turkey's south. This is his "Peace Pipeline," which would carry water from the Seyhan and Ceyhan Rivers as far south as Medina and Mecca in the west and from the Tigris River in Turkey to the UAE in the east (Map 11.2). The international contractors Brown and Root (1988) have already presented an introduction to a feasibility study of this project

Their initial presentation suggests 3.5 Mm^3 per day (1.28 × 10^9 m^3/yr) of water flowing south in the western pipeline (Map 11.2) and 2.5 Mm^3 per day (0.9 x 10^9 m^3/yr) in the east. Various reports give the combined cost of these ventures at between $17 and $20 billion. Technologically feasible, these seemingly expen-

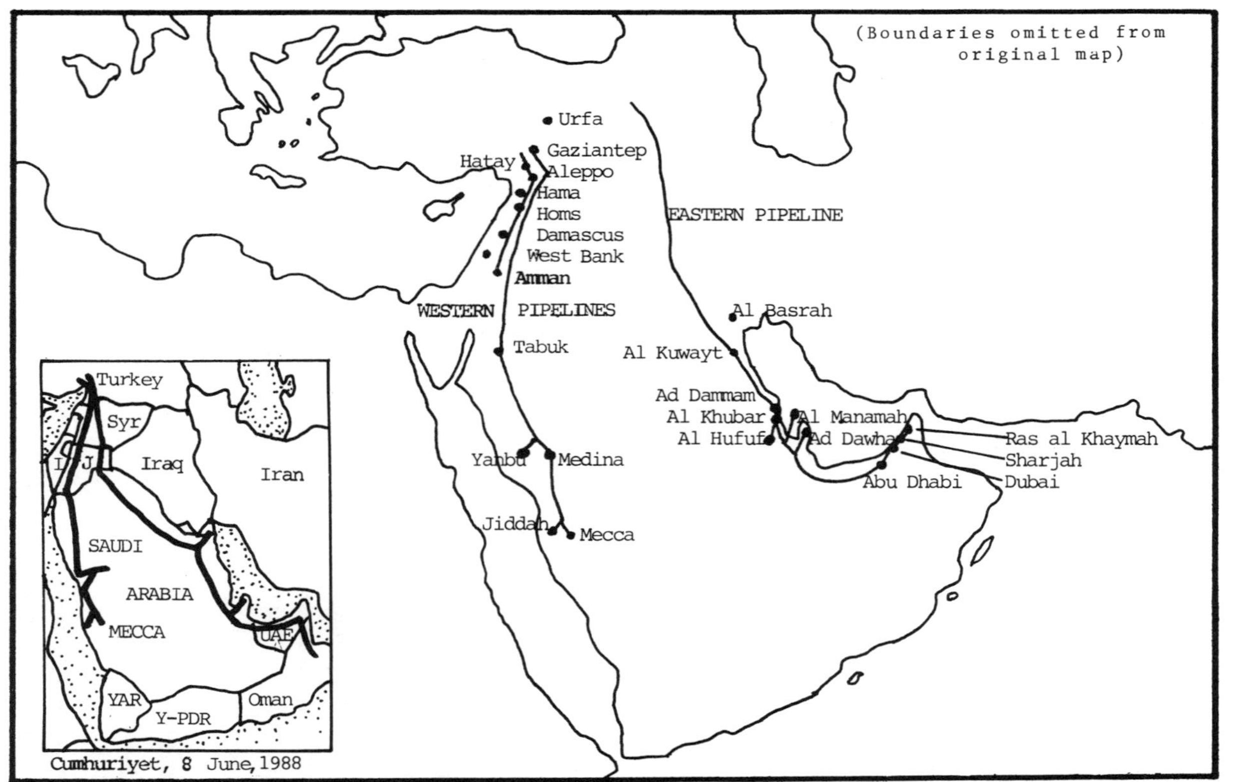

Map 11.2. Alternate routes for the "Peace Pipeline." Source: Abstracted from Brown and Root (1987).

sive lines could deliver water—according to Brown and Root—at one-third the cost of a similar desalinized quantity.

While Premier Ozal did not mention compensating for GAP's depletions, if and when they occur, similar inter-basin transfers originating in Turkey could solve in a reasonable way the problem GAP presents. Map 11.3 shows the distribution of surplus waters throughout Turkey. Each symbol indicates the total amount of surplus after evapotranspiration and other natural losses have been subtracted from estimated precipitation. The empty portion at the top of each symbol is the amount estimated that will be consumed by agricultural and domestic activities. Some river basins will need augmentation. The central, west, and northwest parts of the country are already suffering water shortages. The south and east, however, have ample surpluses capable of satisfying both Turkey's needs and those of its southern neighbors. Some caution must be exercised with this initial, and admittedly rough, estimate. This is indicated by Figure 11.3, which correlates the total surplus remaining in each basin after anticipated depletions with the total available run-off before in-basin use.

The Euphrates and Tigris rivers have sufficient surpluses (according to the data provided by DSI—see Table 11.9), but these must be shared with downstream users, as is the case with the surplus from the Coruh. Any basin falling within the "before and after," less-than-five-billion-cubic-meter category will probably need some supplementation in the future. Those having original flows of between 5 and 10 billion m^3 can probably get by, but have little to offer the others. This would be especially true during runs of dry years. It should be noted that the Seyhan and Ceyhan rivers are in this marginal category, prompting questions about their suitability for use as sources of water for the "Peace Pipeline", particularly in view of the water shortage that may develop in Hatay along the Orontes (Asi) River as dams are built by the Syrians at Zwizun and Kastun (MEED, 11 November, 1988). Much more hope can be attached to the Eastern Mediterranean and Eastern Black Sea Regions. But, in any case, the fact remains that, properly managed, the water resources of Turkey could alleviate much of the water shortage in the Levant and the Peninsula.

How real a possibility is the "Peace Pipeline"? Are the Turks

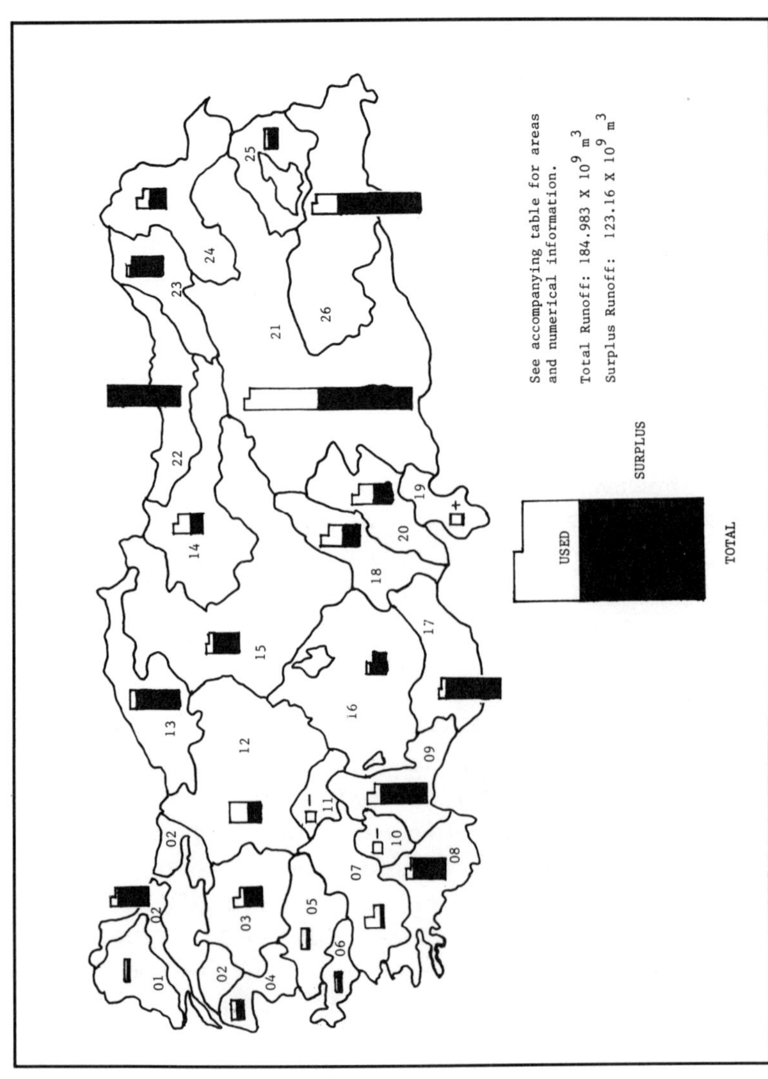

Map 11.3. Total runoff and surplus runoff for 26 drainage basins of Turkey (ca 2000+). Source: DSI (1984b); computations by Kolars.

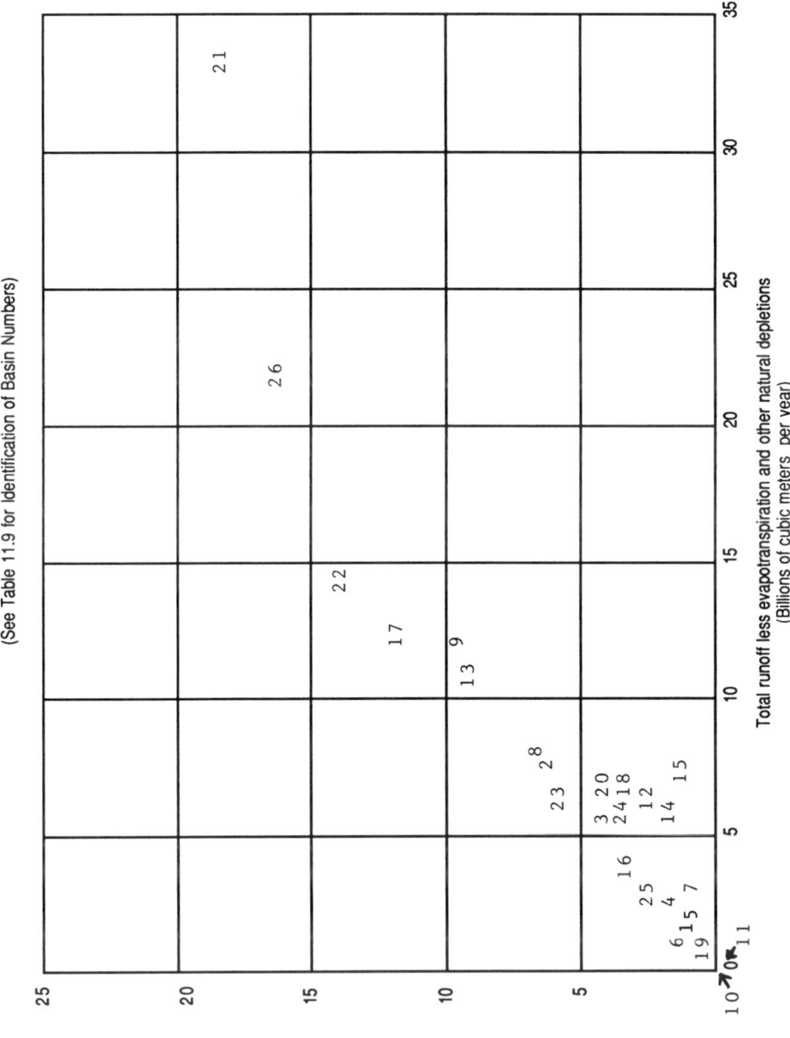

Figure 11.3. Total runoff and surplus runoff of the 26 drainage basins of Turkey.

Table 11.9. Total Runoff and Surplus Runoff of 26 Drainage Basins in Turkey, ca 2000+ (in Mm^3)

No.	Name	Total Runoff	Maximum for Irrigation*	Domestic Use	Total Use	Surplus Runoff	Surplus as % of Total
1	Meric	1,250	631.64	60.0	691.6	558	44.6
2	Marmara	7,620	500.23	829.0	1,329.2	6,291	82.6
3	Susurluk	5,350	1,704.74	147.2	1,851.9	3,498	65.3
4	Kuzey Ege (N Aegean)	2,200	839.39	18.0	857.4	3,498	61.0
5	Gediz	1,810	1,308.30	0.5	1,308.8	501	27.7
6	Kucuk Menderes (Little Menderes)	1,120	31.50	128.0	159.5	961	85.8
7	Buyuk Menderes (Big Menderes)	2,950	2,326.65	—	2,326.7	623	21.1
8	Bati Akdeniz (W Mediterranean)	7,760	1,173.92	—	1,173.9	6,586	84.9
9	Antalya	11,240	1,928.79	—	1,928.8	9,311	82.8
10	Burdur	310	312.16	—	312.2	−2	−0.6
11	Akarcay	450	748.3	—	748.3	−298	−66.2
12	Sakarya	6,030	2,664.13	822.0	3,486.1	2,544	42.2
13	Bati Karadeniz (W Black Sea)	10,040	951.68	18.6	970.3	9,070	90.3
14	Yesil Irmak	5,540	3,261.18	127.0	3,389.0	2,151	38.8
15	Kizil Irmak	6,200	4,973.20	96.0	5,079.2	1,121	18.1

16	Konya	3,360	2,439.24	123.9	263.1	3,097	92.2
17	Dogu Akdeniz						
	(E. Mediterranean)	12,270	712.13	66.6	778.7	11,491	93.7
18	Seyhan	7,060	3,426.33	—	3,426.3	3,634	51.5
19	Asi (Orontes)	1,200	1,165.76	—	1,165.8	34	2.8
20	Ceyhan	7,210	3,615.48	—	3,615.5	3,594	49.8
21	First (Euphrates)	33,480	15,068.67	82.5	15,151.2	18,329	54.7
22	Dogu Karadeniz						
	(E Black Sea)	14,000	9.24	—	9.2	13,991	99.9
23	Coruh	6,460	364.17	—	364.2	6,096	94.4
24	Aras	5,540	2,863.52	—	2,863.5	2,676	48.3
25	Van	2,590	676.41	—	676.4	1,914	73.7
26	Dicle (Tigris)	21,810	5,253.36	—	5,253.4	16,557	75.9
	TOTAL	184,930	59,250.32	2,520.1	61,770.4	123,160	66.6

Source: DSI (1984b, Chapter 5).

Values in columns 2–6 computed by this author.
*DSI gives two figures for irrigable land. The smaller figure under "Benefits after development" has been used to compute these values. An average water loss of 1 m^3/m^2 has been taken as average depletion for Turkey, a figure based on computations for the Euphrates river basin by this author.

serious and what's in it for them? An unusual insight may be gained from the writings of Korkut Ozal, the younger brother of Turgut Ozal and co-author with him of a paper, "On the principles and methods of hydroelectric development planning" (1965), although reference here is made to an independent article by Korkut Ozal et al., "Development of the Euphrates basin in Turkey—a case study" (1967). One wonders if he had not consulted his older brother before committing his ideas to print. He writes:

> Generally speaking, cooperation between the countries within an international river basin is indispensable to resolve conflicts and to increase economic efficiency of resource utilization. . . . The settlement of international water disputes cannot be made according to international law. Such law does not and cannot exist. It becomes therefore the responsibility of the concerned countries to develop a solution to their own problems. This is usually a delicate task, since it is an operation of finding a point of compromise for many diverging claims and demands which normally have their origin in controversies other than the one under discussion. These other controversies may even involve issues which may be regarded, by concerned countries, as issues pertaining to national prestige. Under such difficult conditions, speculative attempts to resolve all these conflicts in a single negotiation are bound to fail. . . . However, those failures should not be considered as indications of the impossibility of cooperation in international basins. The Turkish experience gained on its international rivers with the USSR and Greece suggests that, by a carefully planned sequential approach to the problem, constructive and successful cooperation can be achieved. . . .

There is a strong element of hope and rationality in these words.

What's in all this for the Turks? By making life more secure for their southern neighbors, they would be able to secure, in turn, their southern flank. This would make GAP and its ambitious goals more attainable as well as allowing the redirection of more resources to domestic affairs. It might even mean that, through closer relations with the Syrians and the Iraqis—not to mention the Jordanians and the Saudis—Turkey would be in a position to act as an intermediary between various interested groups, not excluding the Americans, who would find it to their advantage to accept a more quid pro quo relationship with their NATO ally.

And could the Arab nations possibly accept being dependent upon the Turks? First, the water supplied would be a supplement, not the entire supply of any country. Moreover, the Turks have long been dependent upon the Arabs for petroleum and for the loans to buy that much needed energy. It is most unlikely that Turkey would be seen as having the same ambitious intent for the entire Middle East as might the Soviet Union or the United States.

Speculation on such matters is the realm of the political scientist and the diplomat. However, given the growing thirst of the southern lands, the water to the north, and an increasingly undeniable need for cooperation among all the countries involved in such a "Peace Pipeline", it is quite conceivable that we will see in our lifetime a *Pax Aquarum* in this part of the Middle East, one in which Turkey will play an important role.

Appendixes

References

Index

Appendix A:
System Efficiency and Return Flow

System efficiency is defined as that proportion of water removed from a river or reservoir that actually serves the evapotranspiration needs of a particular crop in a particular field. It may be further divided into *delivery efficiency*—the corollary of which is *conveyance loss* (depletions from spills, leakage, evaporation from canal surfaces, management mistakes, misdirections, etc.)—and *on-farm efficiency*—the corollary of which is on-farm loss (spills, leaks, over-irrigation, evapotranspiration by weeds, deep percolation, etc).

Table A.1 lists a number of values for the overall efficiency of farm irrigation systems in the Middle East. This *system efficiency* reflects both delivery efficiency and on-farm efficiency. A second term is introduced, the *coefficient of efficiency*—i.e., the factor by which the calculated water needs of a given crop which are left unsatisfied either by precipitation and/or soil moisture must be multiplied to ensure a sufficient amount of water's being diverted from river or reservoir so that after all losses have occurred the water needed for healthy plant growth will arrive at the plant.

Efficiency ranges from 69.8 percent (coefficient of efficiency: 1.43) to 35 percent (coefficient of efficiency: 2.86). The former value is cited by Waterbury in his *Hydropolitics of the Nile* (1979) as that given by the Egyptian Ministry of Irrigation and is con-

Table A.1. System Efficiency in Near Eastern Irrigation Systems

Value or % Efficiency	Coefficient of Efficiency	Source	Comments
55	1.82	Samman, 24, Euphrates Project	Conveyance loss = 10%; "Field efficiency from water course head regulator" = 60%.
46**	2.17	USAID, II-24, "Syria"	"54% of water diverted into the Homs-Hama canal is lost before it reaches the farmer."
69.8	1.43	Ministry of Irrigation, Egypt: Waterbury, 219	Conveyance loss = 14%; on-farm loss = 16.2%.
35.0	2.86	Conservative estimate for Adapazari Plain, Karataban, 474-475.	"Delivery Eff. = 70%"; "Farm Eff. = 50%".
50.8	1.97	USAID, Egypt, as in Waterbury, Table 23	Conveyance loss = 21.9%; on-farm loss = 27.3%.
0.0	—	Qasim in Khayyat interview	Rasafah area #4 of Syrian Euphrates project (abandoned)

Source: Samman (1980); USAID (1980); Waterbury (1979); Khayyat (1983); Karataban (1966).

*Coefficient of efficiency: the factor by which the amount of water needed for irrigation must be multiplied to indicate the amount needed to be withdrawn from the original source.

**Does not include on-farm loss.

trasted by him with another value arrived at by a USAID team (USAID 1980:I-V) of 50.8 percent (coefficient of efficiency: 1.97). The Ministry's figure is considered to be far too optimistic. In both these cases, the unusual field conditions in the Nile Valley preclude the use of these values in calculations for either Turkey or Syria (Table A.2).

The highest coefficient of efficiency (the least efficient sys-

Table A.2.   Water Use Efficiencies in Egypt

Situation	Lowland	Lowland	Highland	Highland	Highland
Soil Type (carrier)	Clay	Sand	Sand	Sand	Sand
Condition of Carrier (main canal to farm/ farm to field)	Unlined	Unlined	Unlined	Lined	Piped
Main Canal to Farm	0.85	0.80	0.80	0.90	0.90
Farm to Field	0.85	0.80	0.80	0.90	0.90
Soil Type (in field)	Clay	Clay	Sand	Sand	Sand
Method of Application	Surface	Surface	Surface	Sprinkler	Drip
Plant Use	0.65	0.65	0.55	0.70	0.90
Efficiency	0.47	0.42	0.35	0.57	0.73
Coefficient of Efficiency*	2.13	2.38	2.86	1.75	1.37

Source: After Huntington Technical Services as in Allan (1985, 81).

*Coefficient of efficiency: the factor by which the amount of water needed for irrigation must be multiplied to derive the amount needed to be withdrawn from the original source. This has been computed for this study.

tem) is that given by Karataban (1966) for the Adapazari Plain. Karataban's work is considered to be especially pertinent for this discussion because of his taking into account social, economic, and technical conditions in Turkey as well as climatological and phytological considerations. Nevertheless, when this high value is contrasted with two values from Syria and also the full range of efficiencies for Egypt, Karataban's value seems somewhat pessimistic. Tables A.3 and A.4 show what might be considered a "most expensive" case (in terms of water consumed) for the Urfa-Harran and Mardin-Ceylanpinar portions of the GAP irrigation scheme. In these cases Karataban's efficiency values have been combined with total evapotranspiration needs (as given in GAP 1980: III-36) for 136,000 and 206,000 ha respectively.

At this point a second definition must be given, that is, the amount of *return flow*—defined as that portion of the water removed from river or reservoir to meet irrigation needs that finds its way back into the system. It should be noted that water in addition to that needed for evapotranspiration must be provided

Table A.3. Urfa-Harran Water Use—Urfa Tunnel
(136,000 ha per GAP)

Month	I[a] Evapotrans. (m³/ha/mo)	II Total Project Demand (Mm³) (I × area)	III[b,c] Amt Delivered to Fields (Mm³) (II × 2.88)	IV[d] Amt Returned to Streams (Mm³) (34% of III)	V Total Deficit (Mm³) (66% of III)
April	337.09	45.84	132.00	lag	87.11
May	572.36	77.84	224.18	44.89	147.96
June	1694.31	230.43	663.63	76.22	438.00
July	2654.71	361.04	1039.80	225.63	686.27
August	2324.13	316.08	910.32	353.53	600.81
September	1140.81	155.15	446.83	309.51	294.91
October	196.67	26.75	77.03	151.92	50.84
Nov	—	—	—	26.19	—
TOTAL	8920.08	1213.10	3137.80	1143.90	2305.90

Source: GAP (1980, V-4,5).

[a]Equivalent to mm standing water. For example, "Total" is equivalent to 892 mm water per m².
[b]According to "conservative" estimates by Karataban (1966), "Farm efficiency" in Turkey = 50% and "delivery efficiency" = 70%. In order to deliver a required amount to the plants (col. II), the evapotranspiration need must be multiplied by 2.88 (i.e., col. III less 70%, less 50% = amount in col. II).
[c]The terms "farm efficiency" and "delivery efficiency" assume sufficient rain during the wet season to wash away accumulated salts.
[d]Kilic et al. (1966, 70) gives a value of 17/50th, i.e., 34%, for return of water from irrigated fields to streams for all Turkey. Al-Hadithi (1978, 245 and 78, Table 14) uses values of approximately 25% and 30%. FAO (1966, 79) allows "30% to 40%" for return to streams from fields.

Table A.4. Mardin-Ceylanpinar Water Use (est.)—Hilvan Pumpage (206,000 ha per GAP)

Month	I Evapotrans. (m³/ha/mo)	II Total Project Demand (Mm³) (I × area)	III Amt Delivered to Fields (Mm³) (II × 2.88)	IV Amt Returned to Streams (Mm³) (34% of III)	V Total Deficit (Mm³) (66% of III)
April	505.34	83.50	240.48	lag	158.72
May	832.87	171.57	494.13	81.76	326.13
June	2090.56	430.66	1240.29	168.00	818.59
July	2890.21	598.38	1714.70	421.70	1131.70
August	2438.08	502.24	1446.45	583.00	954.66
September	1169.28	204.87	693.71	491.79	457.85
October	172.37	35.51	102.30	235.86	67.53
Nov	—	—	—	34.77	—
TOTAL	9998.71	2059.73	5932.03	2016.89	3915.18

Source: GAP (1980, V-4,5).

Note: This does not include 60,000 ha to be watered by pumping from local aquifer. See Table A.3 for explanation of entries.

in the fields in order that salts—either from leaching of local soils followed by capillary movement upward and evaporation, or by the evaporation of introduced mineralized waters—can be washed from the soil and carried away from the fields. In some cases it is assumed that on-farm losses will include water sufficient to carry away such salts. On the other hand, in none of the examples found for this analysis was such an inclusion made specific and in some cases it was clear that that increment was not included. Close inspection of the cases listed in Table A.5 suggests that a return flow of 35 percent, while generous, is within realistic limits. This value has been used in Tables A.3 and A.4 and in subsequent calculations.

The results of the calculations given in the above two tables indicate an overall depletion of river water of 6,220 Mm³/yr for the 342,000 ha of land in question. As will be seen in section *Water Use per Hectare* this estimate does not effectively take precipitation and soil moisture into account and also uses Karata-

Table A.5.   Return Flows in Near Eastern Irrigation Systems

Value %	Comments	Source
30	For Euphrates systems in Syria in general	Samman, 24
30–40	To the Khabur at Suwar—the shortening of the period of extreme low flow due to increased return flow was mentioned.	FAO, 79
25	Assumed for all irrigation in Turkey, Syria, and Iraq along the Euphrates	al-Hadithi, 245, Table G-4
30	Turkey and Syria	al-Hadithi, 78, Table 14
34	All of Turkey	Kilic, et al., 70
51	All of Egypt, ca 1977. This high value reflects special conditions of water-logging and presents problems of salination, etc.	Waterbury, 218

Source: Samman (1980); FAO (1966); al-Hadithi (1978); Kilic et al. (1966); Waterbury (1979).

ban's (1966) high coefficient of efficiency (in the case of his actual calculations: 2.88). A return flow of 3,160 Mm^3/yr would issue from the same fields in addition to the 6,220 Mm^3/yr consumed. While these figures will be refined in the summary portion of this analysis, they are included at this point as examples of the range of values possibly used during the planning stages of GAP.

After careful consideration, it was decided to use an overall *system efficiency* of 40 percent (coefficient of efficiency: 2.5) for calculations in this analysis. This lesser value somewhat ameliorates the pessimistic view taken by Karataban (1966) and assumes that the return flow is included in the initial figures calculated for diversions from rivers and reservoirs. Needless to say, all such estimates must be recalculated as more data relating to actual crops, field size, condition of delivery systems, etc., become available. What is offered here is an outline painted with the broadest of brush strokes.

Appendix B:
Natural River Flow

None of the Euphrates records cited represents the natural river flow. In order to estimate the average natural river flow, it is necessary to add the amounts of water diverted to the flow measured at some point below all major tributaries. Using figures from the report of Hathaway et al. (1965)—in which estimates of irrigation diversions in Turkey, Syria, and Iraq are made—it is possible to calculate an estimated natural river flow as illustrated in Table B.1.

Table B.1. Calculations of Average "Natural" Flow of Euphrates River at Hit, Iraq
(millions of m^3)

	Measured River Hit, Iraq	Diversions in Turkey	Diversions in Syria	Total "natural" river at Hit
Jan	1.86	0	0.05	1.91
Feb	1.94	0	0.07	2.01
Mar	3.14	0	0.24	3.38
Apr	5.77	0.07	0.24	6.08
May	6.54	0.17	0.37	7.08
Jun	3.27	0.30	0.46	4.03
Jul	1.48	0.37	0.52	2.37
Aug	0.86	0.33	0.44	1.63
Sep	0.73	0.18	0.29	1.20
Oct	0.90	0.07	0.14	1.11
Nov	1.21	0	0.09	1.30
Dec	1.54	0	0.05	1.59
Total	29.24	1.49	2.96	33.69

Source: Adapted from Hathaway et al. (1965) as given in Clawson et al. (1971, 205).

Note: Diversions were estimated as net diversion after taking into account "return flow" from irrigated lands whose areas and cropping patterns were made available to the authors.

References

AMER. *See* Associates for Middle East Research.

Ab-El-Al, I. ca 1965. Statics and dynamics of water in the Syro-Lebanese limestone massifs. Arid zone program II. Ankara Symposium on Arid Zone Hydrology.

Ahmad, M. U. 1981. The role of the Sahara in food production. *Water International* 6:126–129.

———. 1987. Azraq: The end of an oasis? *The Middle East* no. 148 (Feb): 37.

Allan, J. A. 1985. Irrigated agriculture in the Middle East: The future. In *Agricultural development in the Middle East,* edited by P. Beaumont and K. McLachlan, 51–62. New York: John Wiley.

Aresvik, O. 1975. *The agricultural development of Turkey.* New York: Praeger.

Artukoglu, N. O. 1966. An engineering and economic comparison of some projects for the Lower Firat development (EU). In *Symposium on hydrology and water resources development,* 93–105. Ankara: Central Treaty Organization.

al-Ashhab, K. 1983. The transformation of 150,000 hectares into irrigated land. *Al-Thawrah* (Damascus) 12 March: 5.

Associates for Middle East Research. 1986. *Hydrology of Lebanon: Preliminary technical report.* Philadelphia: Associates for Middle East Research.

Ayyildiz, M., et al. 1986. Irrigation technology to be applied in the GAP. In *G.A.P. Tarimsal Kalkinma Simpozyumu—18–21 Kasim* 1986 (in Turkish), edited by the Agricultural Faculty,

Ankara University, 305–328. Ankara: Ankara University Press.

BBC. *See* British Broadcasting Company.

Barchard, D., J. Whelan, and G. Weston. 1983. *MEED Special Report: Turkey* 27: 1–56.

Beaumont, P. 1978. The Euphrates River—An international problem of water resources development. *Environment Conservation* (Lausanne) 5: 35–43.

Bein, F. L. 1981. Domestic water resource use in rural Sudan. *Ekistics* 291: 430–33.

Bourgey, A. 1979. Tabqa Dam and development of the Euphrates River basin in Syria. *Revue de géographie de Lyon* 49: 343–54.

British Broadcasting Company. 1985. Turkey: Loan for dam project. Ankara (*BBC SWB ME/W/1330/A/S*). London: BBC, 8 March 1985.

Brown and Root. 1987. *Source to consumer.* Houston.

Chesnoff, R. Z. 1988. Water feeds the flames. *U.S. News and World Report* 21 Nov.

Clawson, M., H. H. Landberg, and L. T. Alexander. 1971. *The agricultural potential of the Middle East.* New York: American Elsevier.

Cressey, G. B. 1958. Shatt al-Arab basin—Geological review. *Middle East Journal* 12: 448–60.

Cumhuriyet. 1987a. Samla guvenik ve su protokolul. *Cumhuriyet* (Ankara), 18 Jul.

———. 1987b. *Cumhuriyet* (Ankara), 17–19 Sept.

DSI. *See* Devlet Su Isleri.

Devlet Su Isleri. 1982. *Southeastern Anatolian Project.* Ankara.

———. 1984a. *Dams and hydroelectric power plants in Turkey.* Ankara.

———. 1984b. *1983 statistical bulletin with maps.* Ankara.

———. n.d.a. *Keban Baraji.* Anakara.

———. n.d.b. *Southeast Anatolian Project—Ataturk Dam and hydraulic power plant.* Ankara.

Dubertret, L., and J. Weulerse. 1940. Manuel de geographie: Syrie, Liban et Proche Orient. Beirut: Imprimateur Catholique.

Dunne, T., and L. B. Leopold. 1978. *Water in environmental planning.* New York: W. H. Freeman.

Egypt Ministry of Irrigation. 1981. Industrial water use and wastewater production. *Water Master Plan* 10: i-III–38.

Euromoney. 1984. How Ozal engineers Turkey's recovery: New dam will transform East. *Euromoney* Sept: 6–9.

FAO. *See* U. N. Food and Agriculture Organization.

GAP. *See* Turkish Ministry of Energy and Natural Resources, 1980.

Girgin, I., A. Ones, M. Olgun, N. Sonmez, and A. Balaban. 1986. Yerlesme Duzenlemesi ve Altyapi Sorunlari. In *Turkiye Bilimsel ve Teknik Arastirma Grubu, Guney Anadolu Projesi Tarimsal Kalkina Simpozyumu*, 63–90. Ankara: Ankara University Press.

Gischler, C. 1979. Groundwaters in secondary permeable material. In *Water Resources in the Arab Middle East and North Africa*, 63–72. Cambridge: Middle East and North African Studies Press.

Glueckstern, P., and Y. Kantor. Seawater vs. brackish water desalting—Technology, operating problems and overall economics. *Desalination* (Amsterdam) 44: 51–60.

Graham-Brown, S., and D. Barchard. 1981. Turkey taps the Euphrates' resources. *MEED* 25(17 Jul): 50–52.

al-Hadithi, A. H. 1978. Optimal utilization of the water resources of the Euphrates River in Iraq. Ph.D. diss., University of Arizona.

Hathaway, G. A., H. W. Adams, and G. D. Clyde. 1965. Report on international water problems, Keban Dam—Euphrates River. In *Report to the International Bank for Reconstruction and Development.* (See Clawson et al.)

Hershlag, Z. Y. 1960. *Turkey—An economy in transition.* Leiden: E. K. Brill. Ilgaz, N., and M. Demirtas. 1966. Some climatic features of Ankara. In *Symposium on hydrology and water resources development*, 81–92. Ankara: Central Treaty Organization.

———. 1968. *Turkey—The challenge of growth.* The Hague: Uitgeverig van Keulen.

Ivanov, S. A. 1983. Cooperation with Arab countries in the sphere of water-management construction. *Hidrotekhnika i Melioratsia* 7: 76–77.

Jordan Times. 1984. Hassan—Israel water policy pressurizes occupied Arabs. *Jordan Times* (Amman), 26 February.

Karataban, Yilmaz. 1966. An essay on consumptive use and irrigation diversion requirements. In *Symposium on hydrology and water resources development*, 465–76. Ankara: Central Treaty Organization.

Khayyat, M. 1983. For successful agricultural utilization of Euphrates basin: How do we provide conditions for this success? (in Arabic) *Al-Ba'ath* (Damascus) 24: 5.

Kilic, S., A. Aydin, and A. Mat. 1966. Water resources development and hydrology in Turkey. *In Symposium on hydrology and water resources development,* 65–79. Ankara: Central Treaty Organization.

Kolars, J. 1986a. The hydro-imperative of Turkey's search for energy. *Middle East Journal* 40: 53–67.

———. 1986b. The Southwest Anatolia Project: Will it make Turkey a major food supplier to the Middle East? *Middle East Executive Reports* Sept: 9, 19–23.

———. 1987. *The Euphrates River basin: Analysis and water balance.* Water Issues in the Middle East, vol. 2. Philadelphia: Associates for Middle East Research.

———. 1988. "Hydro-geographic background to the utilization of international rivers in the Middle East." In *Proceedings of the 1986 annual meetings of the American Society of International Law.* Washington, D.C.: American Society of International Law.

Kolars, J., T. Casstevens, and J. Wilson. 1983. On-farm water management in Aegean Turkey—1968–1974. *Aid project impact evaluation report* 50. Washington, D.C.: U. S. Agency for International Development.

Lapkin, K. I., Y. D. Rakhimov, and A. V. Pugachev. 1981. Improvement of water supply reliability and problems of partial diversion of Siberian rivers. *Obshchestvenniye nauki v Uzbekistane* no. 1: 59–62.

MEED. *See Middle East Economic Digest.*

Meliczek, Hans. 1987. Land settlement in the Euphrates basin in Syria. In *Land reform, land settlement and cooperatives,* 109–32. Rome: U.N. Food and Agriculture Organization.

Metral, Francoise. 1987. Perimêtres irrigués d'état sur L'Euphrate syrien: Modes de gestion et politique agricole. In *L'Homme et l'eau en Méditerranée et au Proche Orient,* edited by P. Louis and F. Metral, 111–45. Lyon: GS-Maison de L'Orient.

Micklin, P. P. 1986. The status of the Soviet Union's north-south water transfer projects before their abandonment in 1985–86. *Soviet Geography* 27 no. 5: 287–329.

The Middle East. 1984. Turkey and the Arab world: Seeking new markets. *The Middle East* 115: 73–78.

———. 1987. Azraq: The end of an oasis. *The Middle East* 148: 37.

Middle East Economic Digest. 1981a. Pollution outpaces Amman water supply. *MEED* 25(Mar): 15.

———. 1981b. Doubts surround Ataturk Dam scheme. *MEED* 25(Oct): 40.

———. 1984a. Prospects are good for new oil find. *MEED* 28(3 Feb): 33.

———. 1984b. ATA dismisses doubt on Ataturk Dam. *MEED Special Report: Turkey* 28(Jun): 41–45.

———. 1984c. The Saudi wheat conundrum. *MEED* 28 (Jul): 24–25.

———. 1985a. Energy plans—A race against time. *MEED Special Report—Turkey* 29(Jul): 23.

———. 1985b. Turkey: Power schemes set to move ahead. *MEED* 29(1 Mar): 49–51.

———. 1985c. Moves afoot to cut farm support. *MEED Special Report: Saudi Arabia* 29 (Jul): 78ff.

———. 1986a. Syria—In brief. *MEED* 30(9 Aug): 23–24.

———. 1986b. World Bank loans sought for Lower Euphrates scheme. *MEED* 30(4 Oct): 27.

———. 1987a. Soviet Union starts work on main outfall drain. *MEED* 31(31 Jan): 12.

———. 1987b. Turkey—In brief. *MEED* 31(3 Oct): 29.

———. 1988a. In defense of Saudi Arabia's agricultural program. *MEED* 32 (Jan): 22.

———. 1988b. Agricultural success brings new problems. *MEED Special Report: Saudi Arabia* 32(Apr): 19ff.

Middle East Markets. 1984. Middle East: Talks needed on Euphrates River. *Middle East Markets* 11: 2–3.

Milliyet. 1986. Firatin suyu baris getircek. *Milliyet* Nov 27.

Mitchell, W. A. 1977. Partial recovery and reconstruction after disaster: The lice case. *Mass Emergencies* 2: 233–47.

Modern Power Systems. 1985. Ataturk 2400 Mw hydropower project supplies largest irrigation scheme. *Modern Power Systems* 5: 57–66.

Moroney, M. J. 1951. *Facts from figures.* London: Penguin.

Naff, T., and R. C. Matson, ed. 1984. *Water in the Middle East: Conflict or cooperation?* Boulder, Colo.: Westview Press.

Newsweek. 1982. Turkey: Power for the future. *Newsweek.* 5 April: 36.

Ozal, K., R. Kutan, and F. Adak. 1967. Development of the Euphrates basin in Turkey—A case study. In *Water for peace: Planning and developing,* vol. 6, 100–109. Washington, D.C.: U.S. Government Printing Office.

Ozal, T., and K. Ozal. 1965. On the principles and methods of hydroelectric development planning. Paper presented at symposium sponsored by Economic Commission for Europe. Istanbul.

Ozis, U. 1983. Development plan of the Western Tigris basin in Turkey. *International Journal of Water Resources Development* 4: 344–54.

————. 1987. Ancient water works in Anatolia. *Water Resources Development* 3: 55–62.

Parry, J. 1984. Credit for Turkish dam. *The Middle East* 115: 52.

Pearce, F. 1984. Why Qaddafi's wells may run dry. *New Scientist* Sept: 174.

Pitcher, S. 1974. Syria's Euphrates dam promises rapid agricultural development. *Foreign Agriculture* 51: 14–15.

Raslan, D., and A. Fardawi. 1971. Country report of Syria. *In Salinity seminar—Baghdad: Report of regional seminar on methods of amelioration of saline and waterlogged soils,* 215–20. Rome: U. N. Food and Agricultural Organization.

Reisner, M. 1986. *Cadillac desert: The American West and its disappearing water.* New York: Viking.

Royal Embassy of Saudi Arabia. ca 1987. *Saudi Arabia: Agriculture and water resources.* Washington, D.C.

SAR. *See* Syrian Arab Republic.

SCBS. *See* Syrian Central Bureau of Statistics.

Samman, N. 1980. The Euphrates Dam project in Syria. *Industrial Water Engineering* 17: 23–25.

Sanlaville, P., and J. Metral. 1979. Water, land, and man in the Syrian countryside (in French). *Revue de Géographie de Lyon* 51: 229–40.

Sati, B. 1982. Hasakah development viewed (in Arabic). *Syrian Times* (Damascus) 16 August, 3.

Seale, Patrick. 1988. *Asad: The struggle for the Middle East.* London: Taurus.

Shchukin, E. A. 1967. Water resources of the Syrian Arab Republic and plans for their use. In *Water resources and water management problems of Asian countries,* 149–57. Moscow: Nauka.

Sonmez, N., A. Balaban, and M. M. Karadeniz. ca 1988. *Guneydogu Anadolu Projesi ve Entegre Bolge Kalkinmasi Yaklasimi.* Ankara: Turkiye Bilimsel ve Teknik Arastirma Kurumu/Tarim ve Ormancilik Arastirma Grubu.

Statistika. 1979. Narodnoye khozyaystvo SSSR v 1978g. *Statistika* (Moscow).

Syrian Arab Republic. 1980. Resources for agriculture. *Syrian Arab Republic—Development Prospects and Policies Report* 1: 248–49.

Syrian Central Bureau of Statistics. 1979. *Statistical abstract 1979.* Damascus.

————. 1980. *Statistical abstract 1980.* Damascus.

————. 1981. *Statistical abstract 1981.* Damascus.

————. 1983. *Statistical abstract 1983.* Damascus.

————. 1985. *Statistical abstract 1985.* Damascus.

TDN. *See* Turkish Daily News.

TGM. *See* Topraksu Genel Mudurlugu.

TMRR. *See* Turkish Ministry of Reconstruction and Resettlement.

TSIS. *See* Turkish State Institute of Statistics.

Tanoglu, A., S. Erinc, and E. Tumerlekin. 1961. *Turkiye atlasi.* Istanbul: Milli Egitim Basimivi.

Tansey, G. 1982. Turks press forward with ambitious dam programme. *World Water* March: 22–25.

Thorne, W., and H. B. Peterson. 1967. Salinity in United States water. In *Agriculture and the quality of our environment,* edited by N. C. Brady. Washington, D.C.: American Association for the Advancement of Science.

Thornthwaite, C. W., and J. R. Mather. 1957. *Instructions and tables for computing evapotranspiration and the water balance,* vol. 10, no. 3. Philadelphia: Drexel Institute of Technology.

Thornthwaite, C. W., J. R. Mather, and D. B. Carter. 1958. Three water balance maps of Southwest Asia. *Publications in Climatology* 11 (no. 1). Centerton, N.J.: Laboratory of Climatology.

Tishrin. 1985. Agricultural development. *Tishrin* (Damascus) 18 March.

Topraksu Genel Mudurlugu. 1978. *Turkiye Arazi Varligi: Kullanman, Siniflar, Sorunlar.* Ankara.

To the Point International. 1977. Desalination breakthrough in Israel. *To the Point International* 21: 34.

Treakle, H. C. 1970. Syria dams the Euphrates—Plans giant draught for cropland. In *Foreign Agriculture,* 8–10. Washington, D.C.: U.S. Department of Agriculture, Foreign Agriculture Service.

Turkish Daily News. 1984a. Turkey's hydro-electric potential. *Turkish Daily News* (Ankara) 5 Jan.

———. 1984b. 25 Dams will be completed in '84. *Turkish Daily News* (Ankara) 30 Aug.

———. 1986. Prime Minister Turgut Ozal. In *Turkey—1986 Almanac,* 433–35. Ankara.

———. 1987. Turban rules again: Religion, politics, and headscarves should not mix. *Turkish Daily News* (Ankara) 7 Jan.

———. 1988a. Turkey, Syria, and Iraq to discuss waterways. *Turkish Daily News* (Ankara) 29 Aug.

———. 1988b. Turkey to discuss regional water with Iraq, Syria. *Turkish Daily News* (Ankara) 21 Nov.

———. 1988c. Turkey, Iraq, and Syria meet on use of regional waters. *Turkish Daily News* (Ankara) 22 Nov.

———. 1988d. Tripartite water talks end without concrete results. *Turkish Daily News* (Ankara) 23 Nov.

316

Turkish Ministry of Energy and Natural Resources, DSI. 1980. *Guneydogu Anadolu Projesi.* Ankara.

———. 1985. *Hydroelectric energy, 1983.* Ankara.

Turkish Ministry of Reconstruction and Resettlement, Regional Planning Department. n.d. *Elazig-Keban Region regional economic and social development plan and land use—Brief presentation.* Ankara.

Turkish State Institute of Statistics. 1964. 1964 *Census of Manufacturing and Service Industries.* Ankara.

———. 1982. *1980 Genel Sanayi ve Isyerlen Sayimi—II, Kucuk Imalat Sanay* [1980 census of industry and business establishments]. Ankara.

———. 1983a. *1980 population census of Turkey.* Ankara.

———. 1983b. *Tarimsal Yapi ve Uretim 1982.* Ankara.

———. 1984. *1982 annual manufacturing industry statistics.* Ankara.

———. 1986. *Statistical yearbook of Turkey—1985.* Ankara.

———. 1988. *Statistical yearbook of Turkey—1987.* Ankara.

———. n.d. *Census of population by administrative division, 16–10–75.* Ankara.

Turkish State Institute of Statistics and Prime Ministry. 1988. *Turkiye Istatistik Yilligi—1987.* Ankara.

UNDTCD. *See* U.N. Department of Technical Cooperation for Development.

USAID. *See* U.S. Department of Agriculture/Agency for International Development and Syria State Planning Commission.

USDA. *See* U.S. Department of Agriculture.

USDI. *See* U.S. Department of the Interior.

USDS. *See* U.S. Department of State.

USGS. *See* U.S. Geological Survey.

Ubell, K. 1971. Iraq's water resources. *Nature and Resources* 7: 3–9.

U.N. Department of Technical Cooperation for Development. 1982. Ground water in the Eastern Mediterranean and Western Asia. *Natural Resources—Water Series* 9: 143–70.

U.N. Food and Agriculture Organization. 1966. *Étude des resources en eaux souterraines de la Jézireh syrienne.* Rome.

U.S. Department of Agriculture. 1970. *Irrigation water requirements,* vol. 21. Washington, D.C.: U.S. Department of Agriculture.

U.S. Department of Agriculture/Agency for International Development and Syria State Planning Commission. 1980. *Syria: Agriculture sector assessment,* vol. 1–5.

U.S. Department of the Interior, Bureau of Reclamation. 1987. International water power and dam construction. In *Informa-*

tion Please Almanac—Atlas and Yearbook—1987, 40th ed, 578–80. Boston: Houghton Mifflin.

U.S. Department of State. 1978. Syria: Euphrates basin irrigation maintenance project: Agreement signed at Damascus July 22, 1976. *United States Treaties and Other International Agreements,* vol. 28, pt. 1.

U.S. Geological Survey. n.d. Recent and projected changes in Dead Sea level and effects on mineral production from the Sea. In *Open File Report,* 78–176. Washington, D.C.

Urfa Provincial Government. n.d. *Urfa—Il Yilligi* 1967. Sivas, Turkey: Dagis Matbaasi.

Wall Street Journal. 1985. Libya launches 25 billion dollar project to quench Sahara nation's thirst. *Wall Street Journal* Oct 3.

Waterbury, J. 1979. *Hydropolitics of the Nile Valley.* Syracuse, New York: Syracuse University Press.

Weatherhead, P. J. 1986. How unusual are unusual events? *The American Naturalist* 128:150–54.

Weekly Economic Report. n.d. Turkey: Tender for Tigris dam construction. *Weekly Economic Report* (London).

Wilkinson T. J. 1978. Erosion and sedimentation along the Euphrates Valley in northern Syria. In *The environmental history of the Near and Middle East,* edited by W. C. Rice. New York: Academic Press.

Withers, B., and S. Vipond. 1980. *Irrigation—design and practice,* 2nd ed. Ithaca, N.Y.: Cornell University Press.

World Bank. 1985. *World development report.* New York: Oxford University Press.

Index

320

John F. Kolars, professor of geography and Near Eastern studies at the University of Michigan, received a B.Sc. in geology from the University of Washington. After working for the United States Geological Survey he later earned a Ph.D. in geography (with special work in anthropology and Near Eastern studies) from the University of Chicago. His initial research focused upon village development and agricultural change in Turkey. For the last decade he has specialized in the natural characteristics and human use of international rivers in the Middle East. He is a regular lecturer at the Foreign Service Institute of the Department of State, as well as a consultant for USAID and the Food and Agriculture Organization of the United Nations.

William A. Mitchell, Colonel, United States Army Air Force, has served as professor and Chair of the Department of Geography at the United States Air Force Academy. He earned his Ph.D. at the University of Illinois for his research "Post Earthquake Turkish Villages: An Analysis of Disaster Related Modernization." In addition to his military duties at the Air War College and in the Middle East, he has been a member of the Disaster Response Center at Boulder, Colorado, and the Associates for Middle Eastern Research (AMER). His continuing research and field work has focused upon human response to earthquakes in Turkey and Iran as well as investigations of agricultural land use patterns in the Middle East and the regional geography of the Euphrates and Tigris river basin.